Raspberry Pi
LED
Projects

Robert J Davis II

Raspberry Pi LED Projects

The reader or builder takes all responsibility for the safe construction and operation of the projects that are contained in this book. It is assumed that the reader already has at least a minimal background in programming and in electronics. This book does not go into as much detail as to how to write the software, or how to build the hardware as books written for total beginners.

The projects found in this book are meant to be simple enough for the average electronics hobbyist. The parts counts were kept to a bare minimum that is needed to get the job done. If I have time later on I will write other Raspberry Pi books that will go into LCD and robotics circuits and designs. The basics are given here so that with some more experimenting someone could build more complex circuits.

Each project in this book has a quick explanation, a schematic diagram, and a software listing. This book is only meant to get you started at building some Raspberry Pi based projects. Feel free to modify, improve, or even "play" with the software and hardware. Electronics can be lots of fun and that is what you should do with these projects. Have fun!

Most of the parts to build the projects in this book are available at your local Radio Shack store or through MCM Electronics. If you have lots of patience then you can also find the parts for sale on eBay. Be aware that many of the sellers on eBay are actually located in China and hence it will usually take a long time for your parts to be shipped to you from China.

Some of the code in this book might appear to be a bit simplistic. I have had readers comment that the code should be easier to understand and to follow. So the code has not been compressed as much as possible and is hence a little longer that what is needed. For instance instead of using "6" it might be "0,1,1,0" that is "6" in binary. When you are setting up an array to turn LED's on and off the "1" and "0" is much easier to read.

Table of Contents:

Chapter 1
Introduction to Electronics

In this book we will be using several electronic components. This is only a brief introduction to some of the electronic components that we will be using for the projects that are found in this book. Let us start out with the "Star" of our show – the LED.

LED stands for "Light Emitting Diode". We will be using several LED's in every project in this book. An LED is a "diode" in that current only flows when power is connected up in one direction. LED's usually light up at around 1.6 volts or 3.2 volts. Since our power source is 5 volts we need to add resistors in series with the LED's to keep from damaging the LED's. Resistors resist the flow of electricity. Since an ideal current for a LED is 10 ma or .01 amps and R=E/I, (5-1.6 volts)/.01 which is 3.4/.01 or 340 ohms. So the resistor that we will use should be larger than 340 ohms. For the first few projects we can use 470 or even 1000 ohm resistors.

Here is a picture of some typical LED's. At the top there are some 3 mm LED's, in the middle there are some 5MM LED's and at the bottom there are some square LED's. They come in red, green, yellow and clear. Clear LED's can be multicolor like in the upper right corner or they can be white or they can be just about any other color.

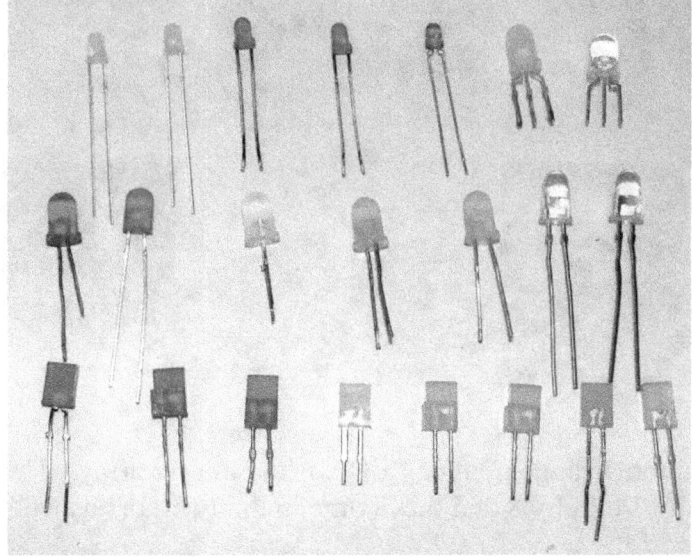

When a LED is new there is one lead that is longer than the other. The longer lead is the positive lead and the shorter lead is the negative lead. There is also a flat spot on the base of the LED next to the negative lead.

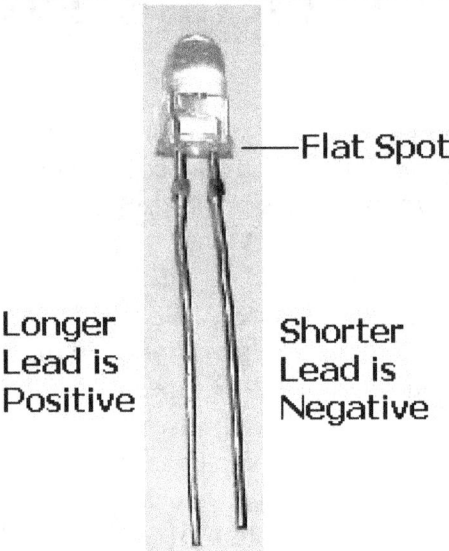

This is the schematic symbol for an individual LED.

LED's also come in arrangements like these seven segment LED's that are used to display numbers.

Another common arrangement of LED's is the matrix array. They come in sizes from 5 by 7 to 8 by 8 and they come in many colors as well.

A "resistor" is a device that resists or limits the flow of electricity. Usually it is made out of carbon, but these days there are lots of things that are used to make resistors. This is what some resistors look like. There are one quarter watt resistors on the left, one half watt resistors on the right and a one watt resistor on the far right.

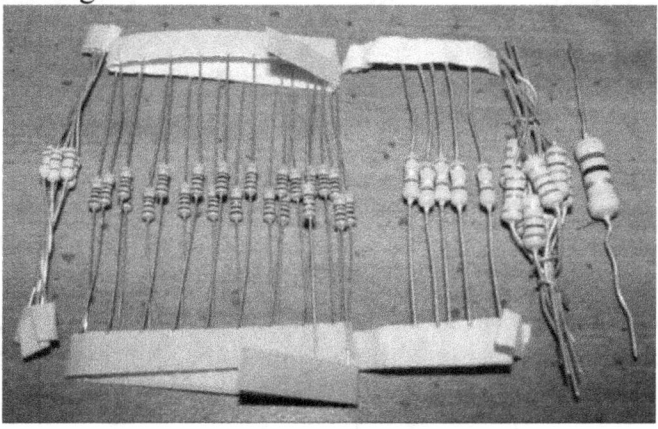

This is the schematic symbol for a resistor.

Transistors are like switches, but they work electronically. They can not only turn things on and off, but they can "amplify" too. That means that they can take a smaller voltage or current in and then turn on or off a larger voltage or current. Here is a picture of some typical transistors. The one on the right is considered to be a "power transistor" in that it can handle a lot of power.

Here are some schematic diagrams of some typical transistors.

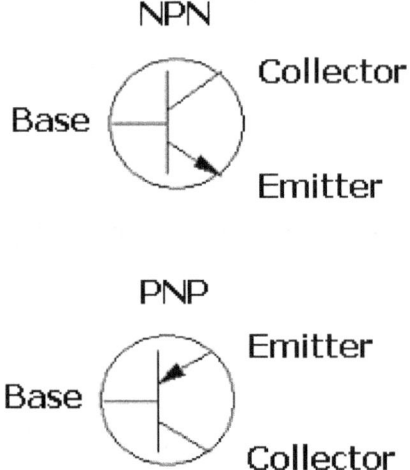

Transistor have three leads, the leads are called the "Base", the "Collector" and the "Emitter". Here is a picture showing how the leads are arranged on a TIP121 power transistor. These power transistors are shown because these same transistors can be used in some of the projects in this book.

Base Collector Emitter

If you put many transistors together into one "case" then you have what is called an "Integrated Circuit" or IC for short. Some have as few as eight transistors, and some have many millions of transistors inside of them. They are very sensitive to static electricity and should always be kept in something that protects them from static electricity.

Here is a picture of some typical IC's.

IC's are usually diagramed as a box with a bunch of wires going in or out of it. Their connections have pin numbers that start with pin 1 at the notched end, and they then go counter clockwise around the IC when viewed from above. Here is the schematic diagram of a 74HC595 IC.

```
Q1 —1        16— 5V
Q2 —2        15— Q0
Q3 —3        14— SER
Q4 —4  74HC  13— OE
Q5 —5  595   12— RCL
Q6 —6        11— CLK
Q7 —7        10— MR
GND —8        9— SOUT
```

A "Breadboard" is a device that allows you to easily and temporally make
and troubleshoot an electronic circuit design. It has several connectors inside
that connect the sets of pins together.

Here is a picture of what a breadboard looks like with the cover removed and
a row of connectors are exposed. There are horizontal rows that the
components are connected on, and there are vertical rows (top to bottom) that
are usually used as the power supply busses.

Another thing you will need is a lot of jumper wires. I make my own
jumpers out of 20 gauge wires. I use three inch and six inch lengths in three
colors. Red is for positive, black or blue for negative and white for signals.
Some more colors would be nice but that is what I have on hand.

I also have some professionally made jumper wires that came in sets of 10
with male plugs on one end and female connectors on the other end. These
jumpers are nice to use for connecting the breadboard projects to the
Raspberry Pi.

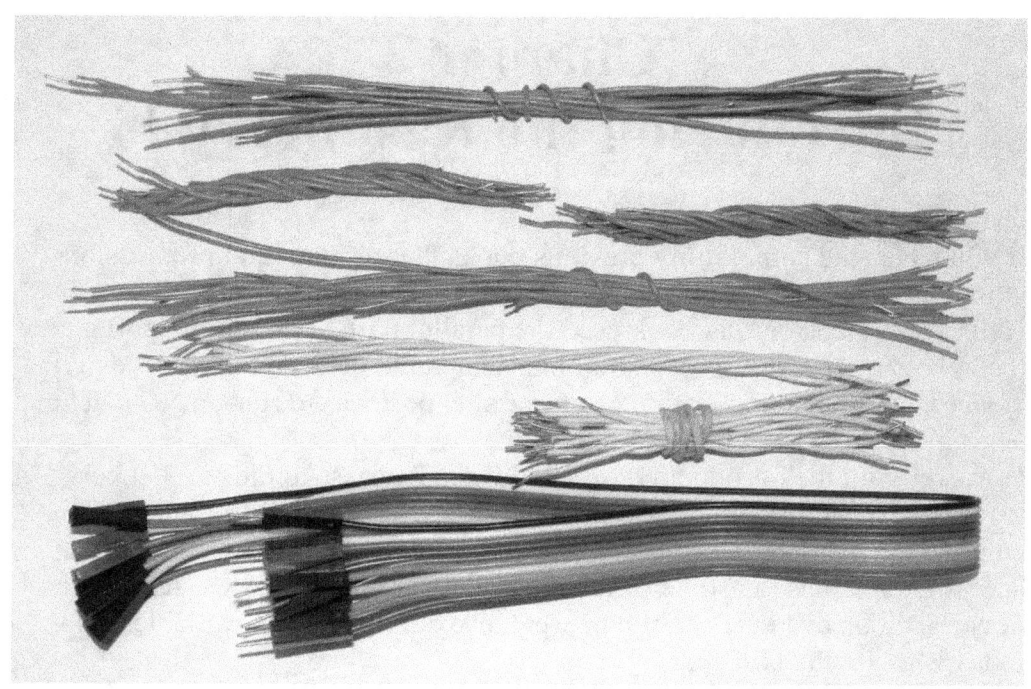

Another thing you can use is an adapter that brings all of the GPIO pins out into a nice neat arrangement. I have one of those adapters but it did not fit any of my breadboards. I have modified it so the power pins come out to some sockets so jumper wires can then be inserted into them.

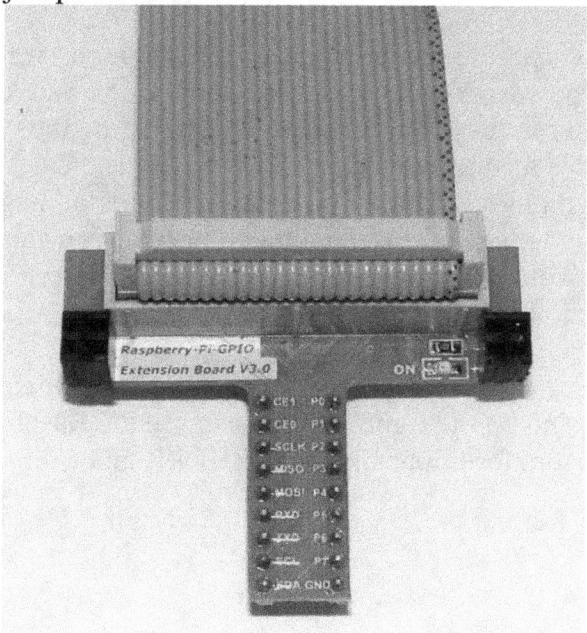

Chapter 2
Introducing the Raspberry Pi

If you have read some of my previous books there is a natural progression going on. My oldest book was "Digital and Computer Projects". The computer projects in that book used the parallel port. The parallel port gave the hobbyist access to some digital pins that could be used to control external devices. One such application of the parallel port was to run a CNC machine.

However with the introduction of Windows XP access to the parallel port was taken away. Eventually the parallel port was removed from the computer altogether. However the serial port gave the hobbyist indirect access to data pins via some serial to parallel converters and via serial devices. Soon even the serial port went away and our only available option today is to use the USB port.

The Arduino Uno can be plugged into any USB port and it gives you 14 digital pins and 6 analog pins to control things with. However you need a PC to develop the software to run on your Arduino. Then you download the software to the Arduino via the USB cable.

The Raspberry Pi is a little bigger than the Arduino Uno. It features a lot more computing power. You can now develop the software on the same computer that you run the software on, thus eliminating the "middle PC". The Raspberry Pi is a computer on a "chip". There is the CPU, Memory, I/O and now even a graphics controller all contained on one Integrated Circuit "chip". The disadvantage of this combination is that you can easily destroy the chip and you would then have to replace the Raspberry Pi. So you have to be careful when playing around with the I/O pins.

This next picture compares the physical sizes of the Raspberry Pi on the left and the Arduino Uno on the right. As you can see the Raspberry Pi has many additional connectors for things like video (HDMI and Composite) and Audio outputs.

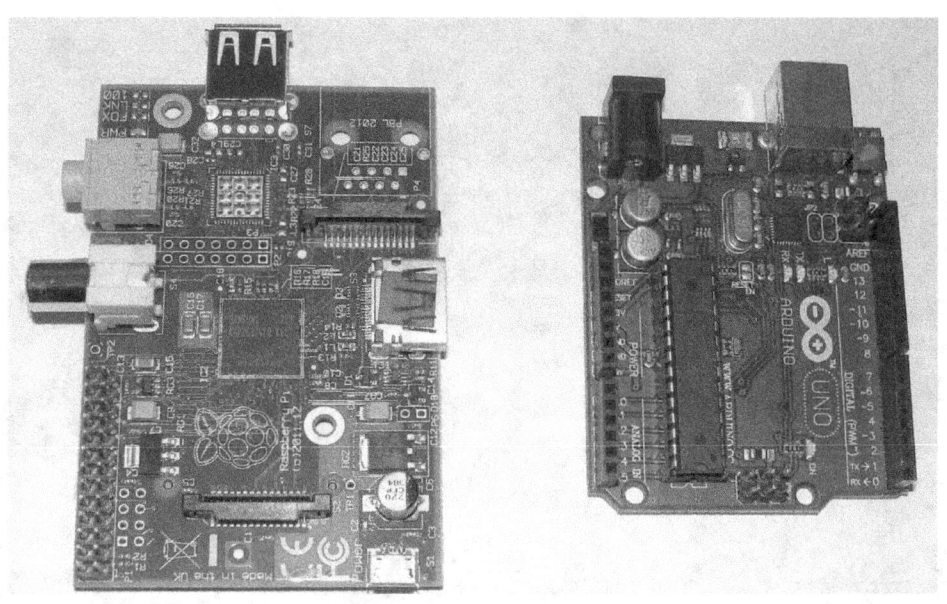

Specification	Arduino Uno	Raspberry Pi
Processor Speed	16 MHz	700 MHz
Memory	32K	256/512 Meg
I/O Pins	14	17/26
Analog Pins	6	None
Power Source	USB or adapter	USB
Video	None	HDMI or Composite
Sound	None	HDMI or sound jack
Memory Jack	None	4 to 16 Gig SD card
Interface Voltage	5 Volts	3.3 Volts

Currently the Raspberry Pi comes in four different flavors.

Spec.	Model A	Model A+	Model B	Model B+
Memory	256Meg	256Meg	512Meg	512Meg
I/O Pins	17	26	17	26
Network	None	None	Yes	Yes
USB jacks	One	One	Two	Two

Up next is a picture comparing the Raspberry Pi Model A on the left and a Model B that is on the right. The biggest visible difference is that the model B has the additional network jack.

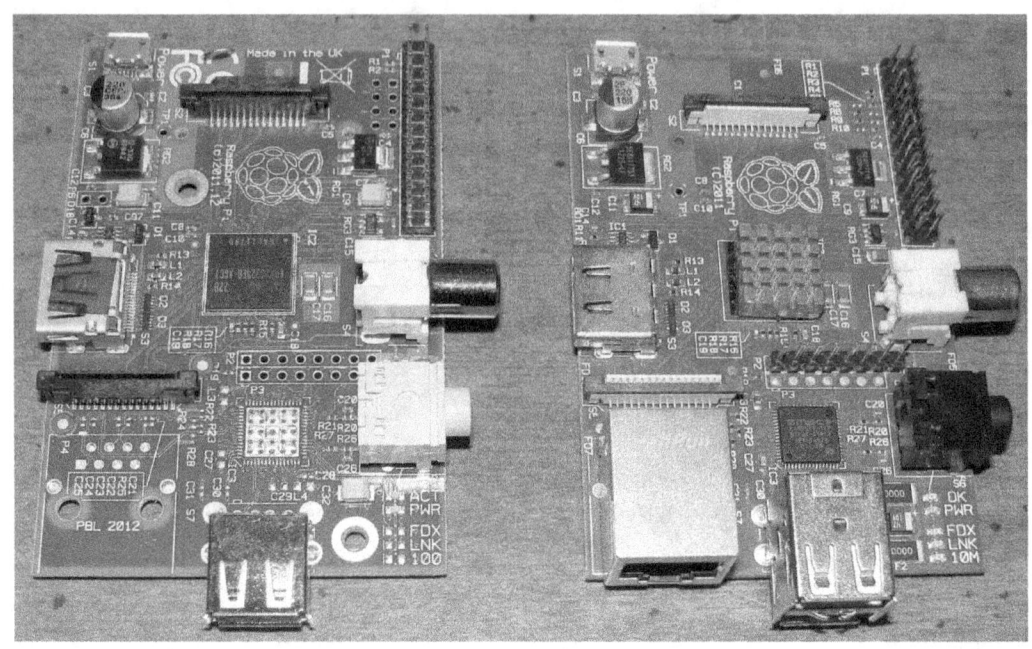

Here is a picture showing the connectors on the Raspberry Pi. There are also two connectors for flexible ribbon cables. They are the camera connector on the right and the display connector on the left.

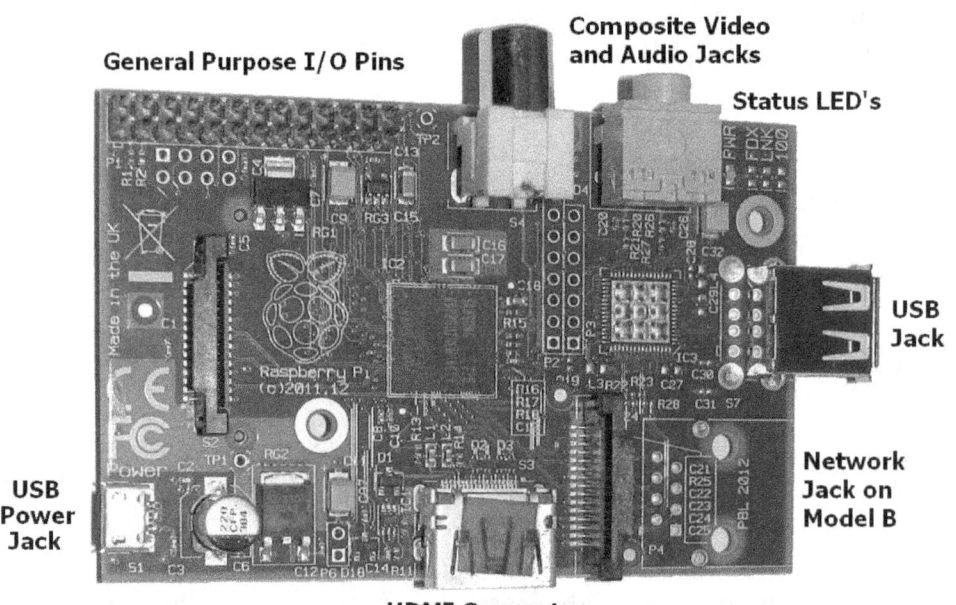

Chapter 3
Setting up the Raspberry Pi

If you expect to be able to use the Raspberry Pi right out of the box you are in for an unpleasant surprise. You will need several things to get it working. However they are things that you might just have lying around some place.

You will need these additional things to get the Raspberry Pi board up and running.

1. A 4, 8, or 16 GB SD memory module.
    It will work with most size adapters.
    Not all SD cards are compatible with the Raspberry Pi.
2. A 5 Volt, preferably 1 Amp (or more) Cell phone charger.
    If using a powered hub you can use a USB to Cell Phone adapter.
3. Preferably a powered USB 2.0 Hub.
    The Raspberry Pi will NOT work with a USB 1.0 Hub.
4. HDMI cable and an HDMI compatible monitor.
    You can use a Composite monitor but the default setting is HDMI.
5. A USB Mouse and a USB Keyboard.
6. An optional USB Wireless Network Adapter.
7. An optional case or something to hold the Raspberry Pi.

The Raspberry Pi has no BIOS (Basic Input/Output System) so it will do NOTHING without a SD card with an operating system installed on it. The easiest way to get started is with "NOOBS" (New Out Of Box System) set up on a SD memory stick. Here is how to set up NOOBS to work with the Raspberry Pi. You can buy SD cards with NOOBS installed, and various other operating systems can be already installed on it.

Up next is a picture of some SD cards that I have used with my Raspberry Pi. They are a 16 GB SanDask, and 8 GB that came with NOOBS on it and a 4 GB that came with an adapter and was used in a camera for a while.

You will need a SD card reader like this one to set up your own SD card.

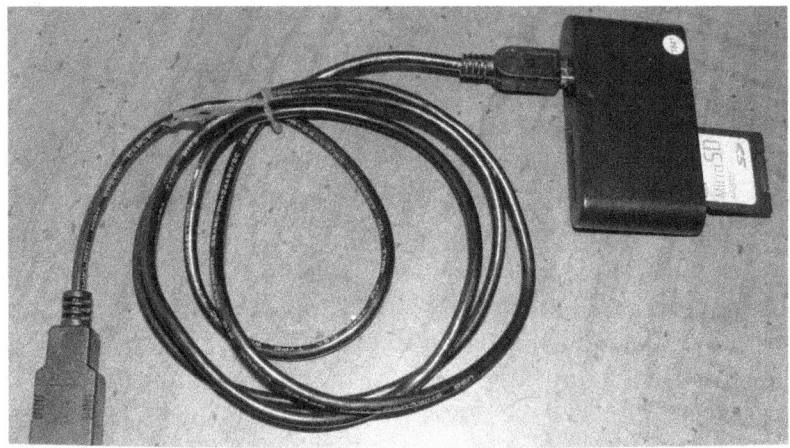

These are the steps you will need to do to set up the SD memory card for use with the Raspberry Pi. Steps 1-3 are done on a PC that is equipped with a SD card reader. Steps 4 and 5 are run on the Raspberry Pi. If you buy a preloaded SD card you only need to do steps 4 and 5.

1. Format the SD memory card.
2. Extract the file "NOOBS_v1_3_11.zip" to a directory.
3. Select all of the NOOBS files and directories and copy them to the SD card that you just formatted.
4. Insert the card into the Raspberry Pi and power it up.
5. NOOBS allows you to select the operating system that you want to install. Select "Raspian [Recommended]" and sit back and wait – it takes a long time. The card will be reformatted to work with Raspbian Linux.

This next picture shows what files should be in the NOOBS directory.

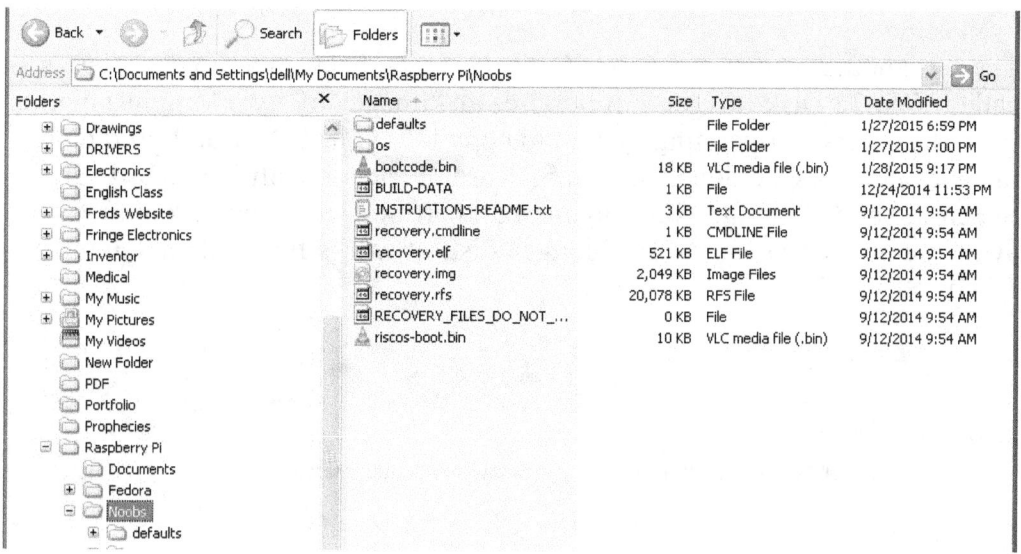

When you boot your Raspberry Pi with NOOBS this screen comes up. Other versions of NOOBS have other available options.

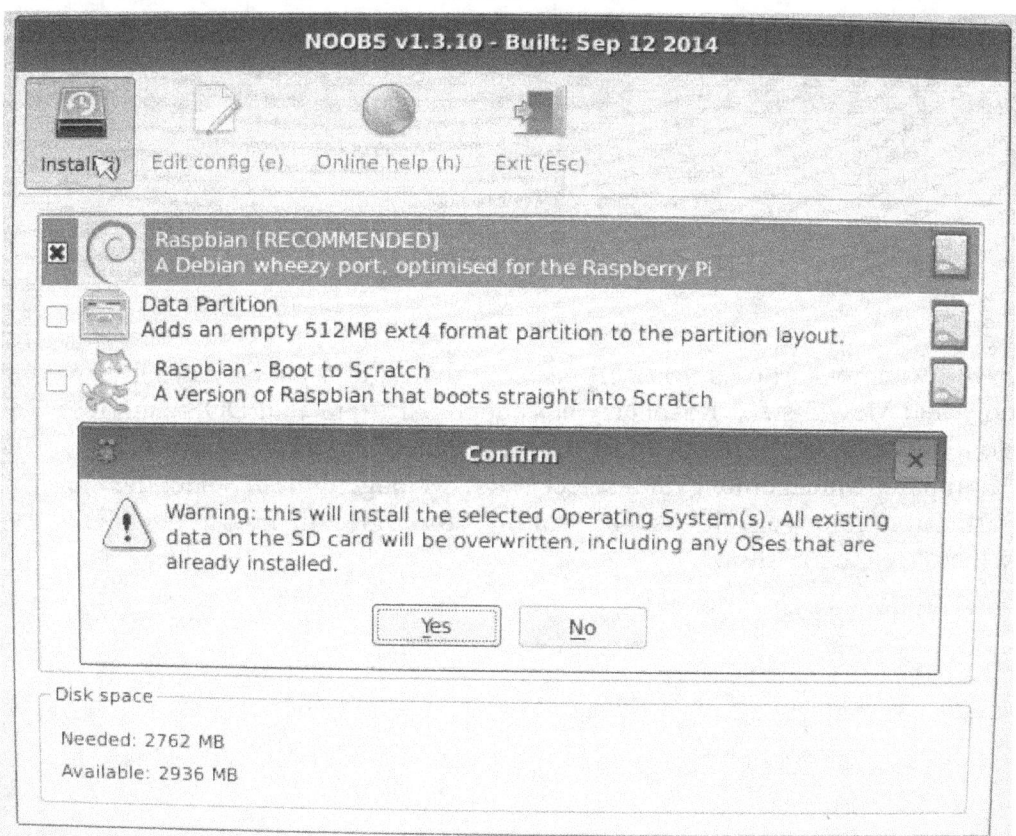

There is another way to get your SD card set up. There is a program that is called "Win32 Disk Imager". It requires an "image" of a working installation for a Raspberry Pi operating system to copy it to your SD card. I used disk imager to install Occidentalist on a SD card. I tried installing it on a 4 Gig SD card, but disk imager kept saying that there was not enough room available on the SD card to fit Occidentalist. It works fine with an 8 Gig or 16 Gig SD card.

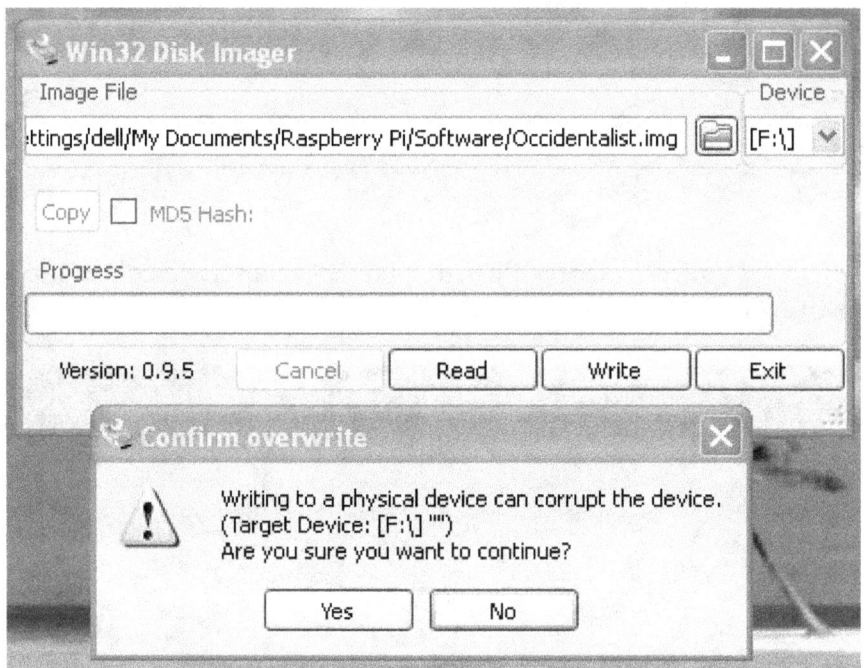

If you want to start over you will discover that your 4 gig SD card is now only 800 Meg in size. Raspberry has added partitions that DOS cannot see. To remove these partitions go to "Control Panel", "Administrative Tools" "Computer Management" and select "Disk Management". Select the partition to remove so it has diagonal lines through it then right click and select "Delete Partition".

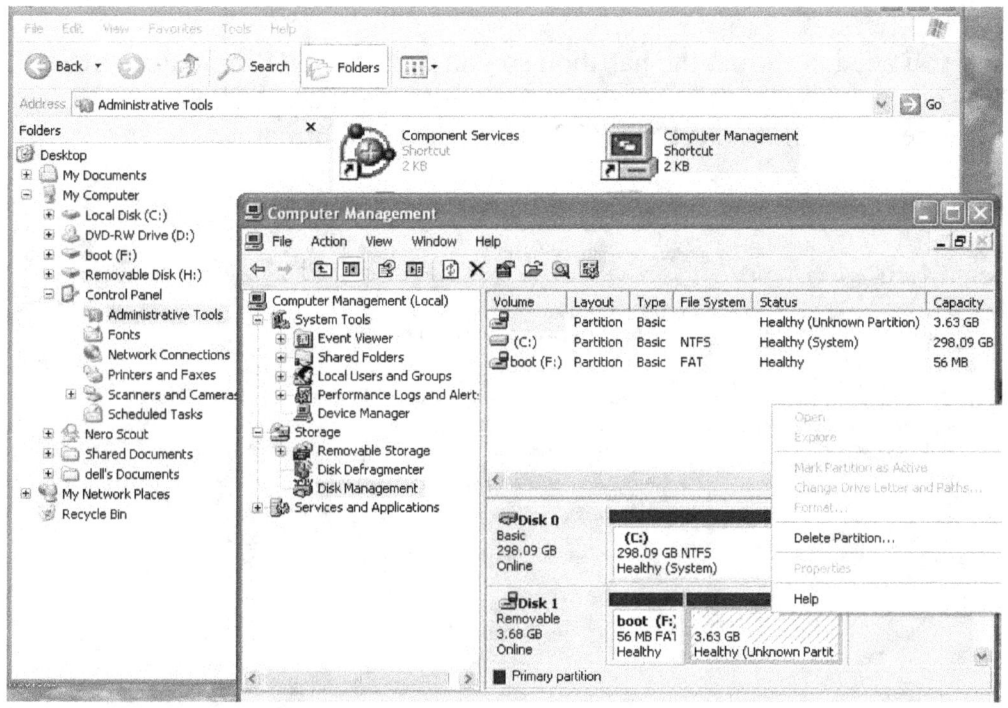

Then you need to create a new partition that includes the entire SD card. Once again select the SD card so that there are diagonal lines through it. Then right click and select "New Partition".

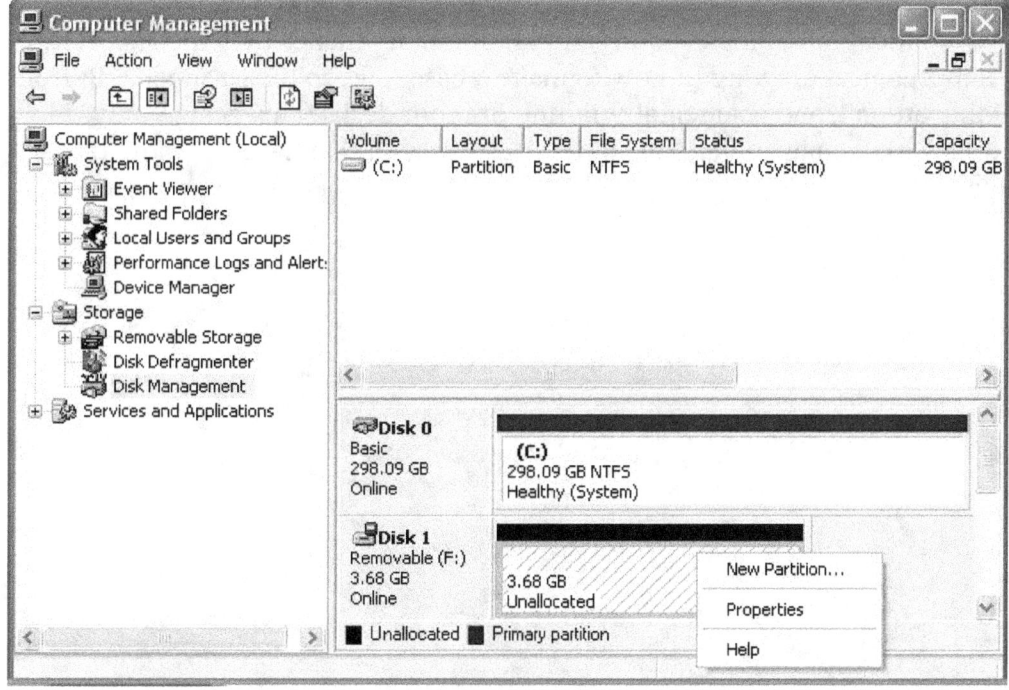

Next you need to format the partition so you can copy NOOBS back onto it. Click on your new partition and select "Format Partition". Click through a few screens and then select "FAT32" and "Next" to format it.

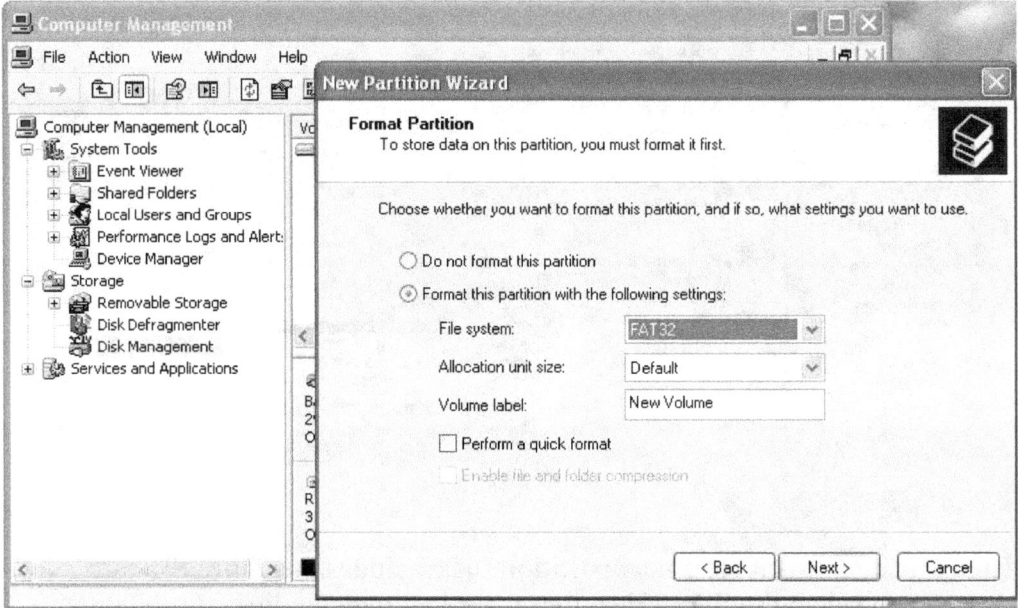

If you have not purchased a case for your Raspberry Pi you can make your own "holder" out of a piece of Plexiglas about 3.5 inches by 5 inches. Here is a mechanical drawing of how to make a base to keep the Raspberry Pi from sliding around, and accidentally falling off your desk. This design only works with the newer versions of the Raspberry Pi that have the mounting holes.

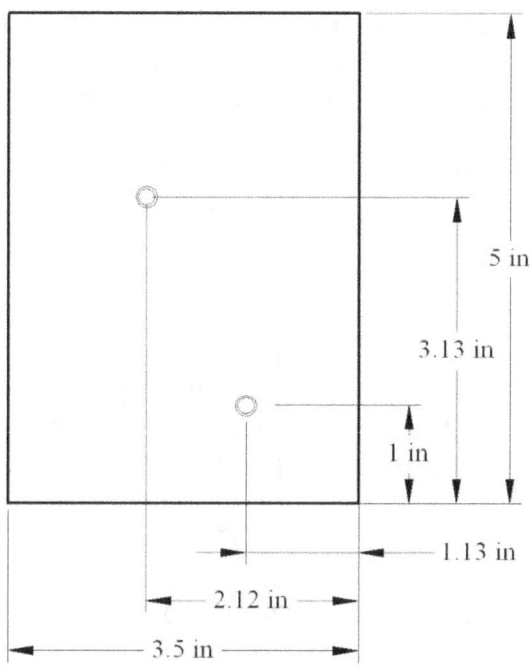

Here is a picture of a Raspberry Pi on the Plexiglas base that was drawn above. I rounded the corners of the Plexiglas with a file.

The Raspberry Pi has 26 GPIO pins many of them have multiple possible definitions. Pin one is at the top left where the board says "P1". Remember that the I/O pins are 3.3 volt pins and they should not be connected to 5 volts!

GPIO – BCM	Name	GPIO – BCM	Name
1. 3.3V		2. 5V	
3. GPIO-0/2	SDA	4. 5V	
5. GPIO-1/3	SCL	6. Gnd	
7. GPIO-4	GPIO7	8. GPIO-14	TX
9. Gnd		10. GPIO-15	RX
11. GPIO-17	GPIO0	12. GPIO-18	GPIO1
13. GPIO-21/27	GPIO2	14. Gnd	
15. GPIO-22	GPIO3	16. GPIO-23	GPIO4
17. 3.3V		18. GPIO-24	GPIO5
19. GPIO-10	MOSI	20. Gnd	
21. GPIO-9	MISO	22. GPIO-25	GPIO6
23. GPIO-11	SCLK	24. GPIO-8	CE0
25. Gnd		26. GPIO-7	CE1

Pin 13 was GPIO 21 on older versions of the Raspberry Pi and is now GPIO 27 on newer versions. The same thing applies to physical pins 3 and 5.

You can purchase an adapter board that neatly organizes the GPIO pins and the adapter usually includes a ribbon cable. This will eliminate the need to connect to the Raspberry Pi's pins directly.

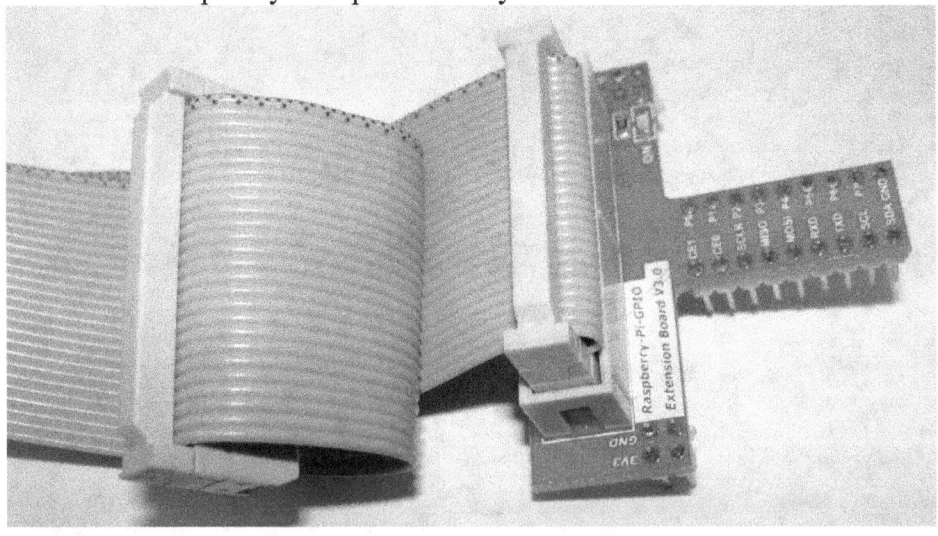

Chapter 4
Introduction to Python Programming

Programming the Raspberry Pi was my first introduction to Python programming. This chapter is written from the viewpoint of someone who does a lot of computer programming and will point out some of the differences. This is not an all inclusive list of commands. This just covers the programming that is used in this book.

The first thing you need to know is the general layout of a Python program. There is no "end if" command, instead the indentation tells when the "if" statement ends. So the indentation is critical, it must remain the same for each statement under an "if" a "while" or a "for". To end an "if" just step back to the indentation position of the last "if"

Case is critical, and lower case is usually used.
Comments always start with the "#" character.
Multiple statements on a single line are separated with ";".
Command lines like "if a==b" end in ":"
    The next lines then say what to do.
Variables do not need to be declared, the value determines the type.
You can assign multiple variables like a=b=c=d=12
Numbers can be int(9), long(1234L), float(-1234.5) and complex.
Strings can be accessed or "sliced" with [1] and [1:5] to pick out characters.
Strings can be strung together with the "+" character.
Lists are enclosed in [] and they are separated by commas.
    List = [123, 345, 456, 678, 789, "test"]
Tuples are enclosed in ( ) and they are separated by commas.
    Tuples cannot be changed once they are created.
Dictionaries are enclosed in { } key:value pairs are separated by commas.
    Dictionary = {'row1':123, 'row2':234, 'row3':345 }

Converts x to a string: str(x)
Converts integer x to a character: chr(x)
Addition: +
Subtraction: -

Multiplication: *
Division: /
Modulus: % (Returns remainder)
Exponent: **
Are they equal: ==
Are they not equal: != or <>
Is it greater than: >
Is it less than: <
Is it less than or equal to: <=
Is it greater than or equal to: >+
To assign: =
Binary AND: &
Binary OR: |
Binary XOR: ^
Binary Left Shift: <<
Binary Right Shift: >>
The if: statement can have an elif: statement.
The while: statement can have an else: statement.
The for: statement can automatically step: or sequence:
      for letter in "test":
Loops can contain statements to break: continue: and pass:
The positive absolute value of x: abs(x)
The logarithm of x: log(x)
The base 10 log of x: log10(x)
The value of x to the power of y: pow(x, y)
To round off to the decimal point: round(x[,n]) n=number of digits.
The square root of x: sqrt(x)
A random item from a list: choice(seq)
A random item from a range: ranrange ([start,] stop [,step])
A random float: random()
Starting value for a random number: seed([x])
Randomize the items in a list: shuffle(list)
Arc cosine of x: acos(x)
Arc sine of x: asin(x)
Arc tangent of x: atan(x)
Cosine of x in radians: cos(x)
Sine of x in radians: sin(x)
Tangent of x in radians: tan(x)

How to create your own Python program.

Writing your first Raspberry Python application can seem a bit daunting at first. You will need to have RPi.GPIO installed on your Raspberry Pi. Then newer versions of Raspian include GPIO. GPIO is also included in Occidentalist.

To create your new program select file manager. Make sure you are in your home directory by clicking on the "home" icon. Right click on a blank area in the right panel and select "Create New" and "Empty File" then name the file something like "led.py".

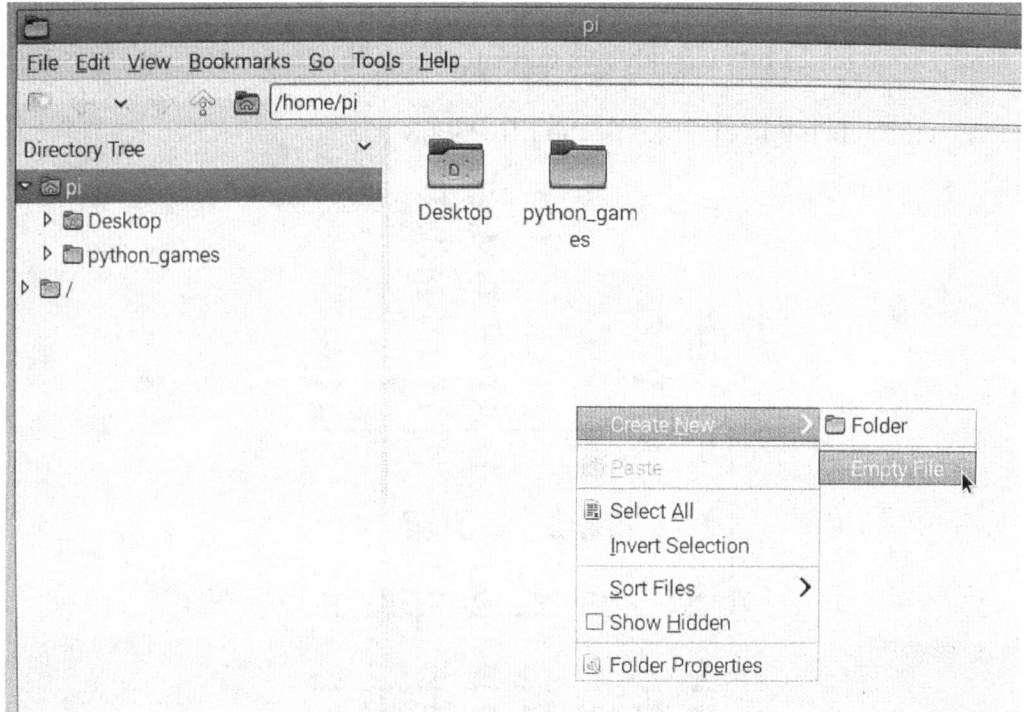

To edit the new "led.py" file just double click on it and it should open in the text editor. Instead of just typing in everything you can use cut and paste to make entering the program much easier. Save you program and look for the "terminal" icon.

From the terminal application type in "sudo python led.py" and press enter. Your Python application should now be running. The most common error messages indicate that you are not in the home directory or that you have a typing error. If your application hangs "Control C" will abort the program.

Chapter 5
Adding Analog Inputs

Unlike the Arduino the Raspberry Pi has no built in analog inputs. This is easily cured by adding a MCP3008 eight port analog to digital converter. It can do 200,000 samples per second at 5 volts but we will be running it at 3.3 volts to keep the input at the Raspberry Pi 3.3 volts logic level. You might be able to go even faster by only using 8 bit conversions. But the need to serially transfer 24 bits will always takes up some time.

Here is the schematic diagram of the MCP3008

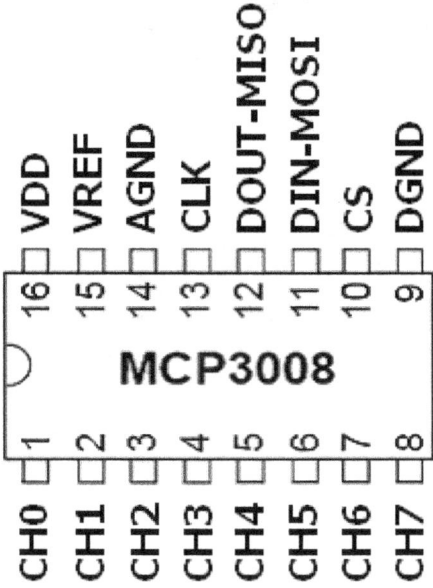

MCP3008	Pi Pin	Signal Name
1-8		Analog inputs 0 to 7
9	6	Gnd
10	24	CE0/8
11	19	MOSI/10
12	21	MISO/9
13	23	SCLK/11
14	6	Gnd
15	1	3.3V
16	1	3.3V

Here is a schematic of how to wire up the MCP3008.

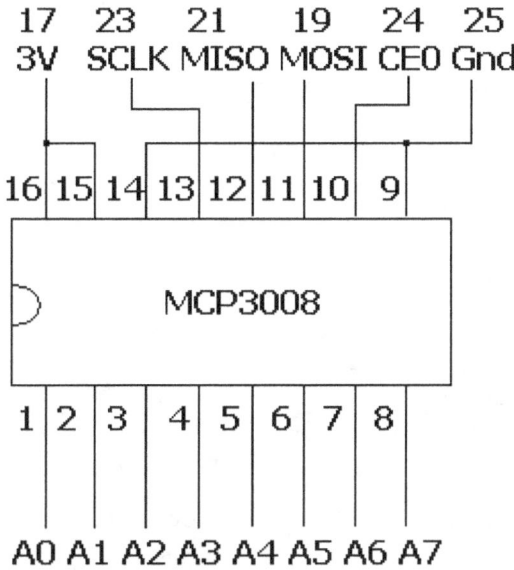

Here is a picture of the MCP3008 wiring. A variable resistor was added.

The communication protocol involves three 8 bit bytes. Going out of the Pi to the MCP3008 the first byte ends in a "1" as a start bit. The next bit is a "1" if you are doing 8 analog converters as opposed to four double ended converters. The next three bits are the analog port selection bits. The remaining 12 bits do not matter as they are ignored.

Serial data starts coming back to the Raspberry Pi at bit 13 to bit 23 for a total of 10 bits. We need to check to see if the input is high or low after each shift then shift the results in "adcin" down one position to reconstruct the analog input value.

Here is the Python code to make it work:

```
# MCP3008 communication
# Prints the contents of all 8 analog inputs.
import RPi.GPIO as GPIO
import time
GPIO.setmode(GPIO.BCM)
GPIO.setup(11, GPIO.OUT) # Sclock
GPIO.setup(9, GPIO.IN) # MISO
GPIO.setup(10, GPIO.OUT) # MOSI
GPIO.setup(8, GPIO.OUT) # CE0
port=0
serout=0x18
while port < 8:
  GPIO.output(8, GPIO.HIGH) # deselect chip
  GPIO.output(11, GPIO.LOW) # set clock low
  shift=0
  adcin=0
  # 24 pits shifted in and out
  while shift < 24:
    GPIO.output(8, GPIO.LOW) # select chip
    # low for most bits out
    GPIO.output(10, GPIO.LOW)
    if (shift == 7 or shift ==8):
      GPIO.output(10, GPIO.HIGH)
    if shift == 9:
      if (port > 3):
        GPIO.output(10, GPIO.HIGH)
    if shift == 10:
      if (port & 0x02):
        GPIO.output(10, GPIO.HIGH)
    if shift == 11:
      if (port & 0x01):
        GPIO.output(10, GPIO.HIGH)
    if shift > 13: # last 10 bits
      if(GPIO.input(9)):
        adcin = adcin+1 #set bit
      adcin = adcin << 1 # left shift 1
    # cycle the clock
    GPIO.output(11, GPIO.LOW)
    GPIO.output(11, GPIO.HIGH)
    shift=shift+1
  print adcin
  port=port+1
# end
```

Chapter 6
LED GPIO Tester

This project can be started with just one LED then expanded up to 15 LED's. I used this project to test the I/O pins on a used Raspberry Pi that I bought on eBay.

Here is the schematic diagram of the LED tester.

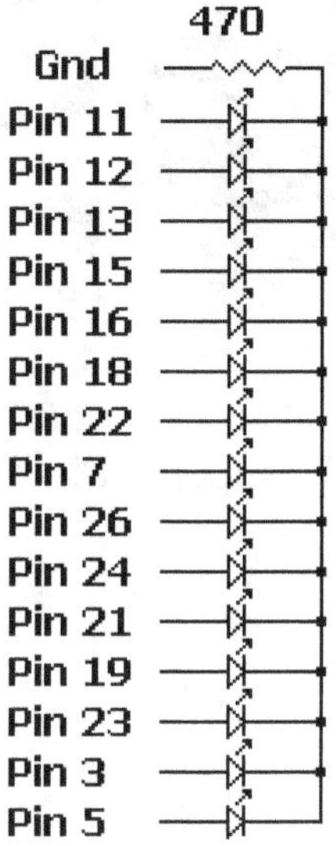

This next picture is of what the 12 LED Raspberry Pi tester setup looks like. The picture was taken when there was only 13 LED's. I later added two more LED's for a total of 15 LED's. There were 5 red, 5 yellow, and 5 green LED's. I used a 10 conductor ribbon cable for most of the connections. There are jumper wires behind the LED's connecting the cathodes together. It took a few tries to get the pins in the correct order so that the LED's light in sequence.

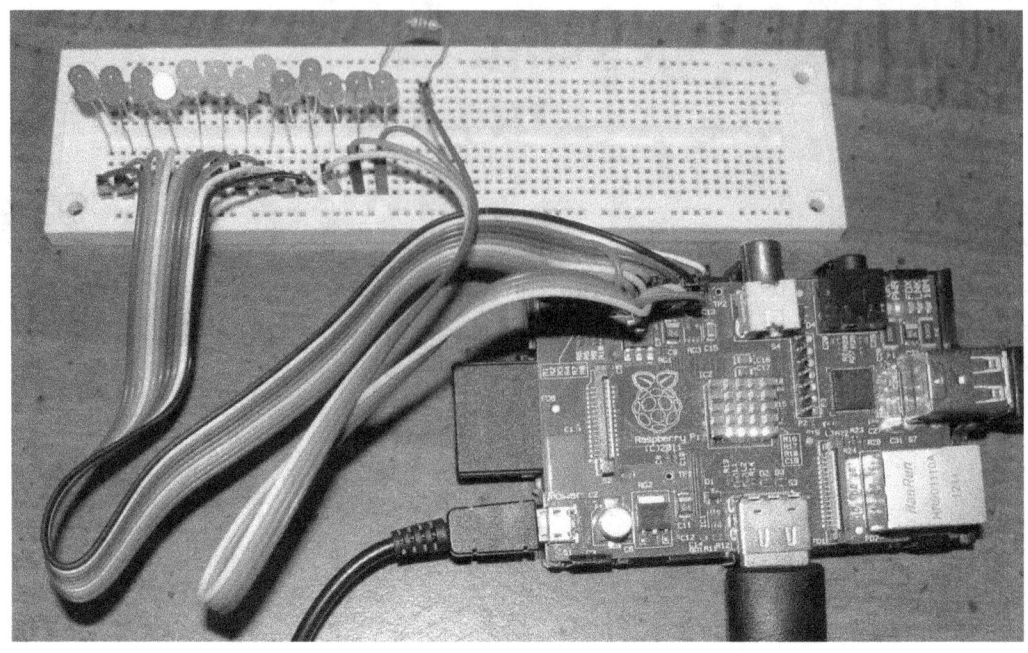

Here is the python code to make it work. Note that pin 21 was replaced with pin 27 in newer versions of the Raspberry Pi.

```
# led.py - GPIO LED tester
# Scans through the LED's connected to the GPIO pins.
import RPi.GPIO as GPIO
import time
GPIO.setmode(GPIO.BCM)
GPIO.setup(17, GPIO.OUT)
GPIO.setup(18, GPIO.OUT)
GPIO.setup(21, GPIO.OUT)
GPIO.setup(22, GPIO.OUT)
GPIO.setup(23, GPIO.OUT)
GPIO.setup(24, GPIO.OUT)
GPIO.setup(25, GPIO.OUT)
GPIO.setup(4, GPIO.OUT)
GPIO.setup(7, GPIO.OUT)
GPIO.setup(8, GPIO.OUT)
GPIO.setup(9, GPIO.OUT)
GPIO.setup(10, GPIO.OUT)
GPIO.setup(11, GPIO.OUT)

while True:
    GPIO.output(17, GPIO.HIGH)
    time.sleep(.1)
    GPIO.output(17, GPIO.LOW)
    GPIO.output(18, GPIO.HIGH)
    time.sleep(.1)
```

```
GPIO.output(18, GPIO.LOW)
GPIO.output(21, GPIO.HIGH)
time.sleep(.1)
GPIO.output(21, GPIO.LOW)
GPIO.output(22, GPIO.HIGH)
time.sleep(.1)
GPIO.output(22, GPIO.LOW)
GPIO.output(23, GPIO.HIGH)
time.sleep(.1)
GPIO.output(23, GPIO.LOW)
GPIO.output(24, GPIO.HIGH)
time.sleep(.1)
GPIO.output(24, GPIO.LOW)
GPIO.output(25, GPIO.HIGH)
time.sleep(.1)
GPIO.output(25, GPIO.LOW)
GPIO.output(4, GPIO.HIGH)
time.sleep(.1)
GPIO.output(4, GPIO.LOW)
GPIO.output(7, GPIO.HIGH)
time.sleep(.1)
GPIO.output(7, GPIO.LOW)
GPIO.output(8, GPIO.HIGH)
time.sleep(.1)
GPIO.output(8, GPIO.LOW)
GPIO.output(9, GPIO.HIGH)
time.sleep(.1)
GPIO.output(9, GPIO.LOW)
GPIO.output(10, GPIO.HIGH)
time.sleep(.1)
GPIO.output(10, GPIO.LOW)
GPIO.output(11, GPIO.HIGH)
time.sleep(.1)
GPIO.output(11, GPIO.LOW)
```

This next Python program makes the 15 LED's go back and forth. This is
done by having two counts that will turn on each LED. This program uses a
variable called "count" and "if" statements to select what LED to turn on. As
a result the program is shorter that the one above.

```
# led2.py - LED scanner
# Scans back and forth through the GPIO pins.
import RPi.GPIO as GPIO
import time
GPIO.setmode(GPIO.BCM)
GPIO.setup(17, GPIO.OUT)
GPIO.setup(18, GPIO.OUT)
GPIO.setup(21, GPIO.OUT)
GPIO.setup(22, GPIO.OUT)
```

```python
GPIO.setup(23, GPIO.OUT)
GPIO.setup(24, GPIO.OUT)
GPIO.setup(25, GPIO.OUT)
GPIO.setup(4, GPIO.OUT)
GPIO.setup(0, GPIO.OUT)
GPIO.setup(1, GPIO.OUT)
GPIO.setup(7, GPIO.OUT)
GPIO.setup(8, GPIO.OUT)
GPIO.setup(9, GPIO.OUT)
GPIO.setup(10, GPIO.OUT)
GPIO.setup(11, GPIO.OUT)
# Set up loop
count=0
while count < 30:
  count=count+1
  #Clear all LED's
  GPIO.output(17, GPIO.LOW)
  GPIO.output(18, GPIO.LOW)
  GPIO.output(21, GPIO.LOW)
  GPIO.output(22, GPIO.LOW)
  GPIO.output(23, GPIO.LOW)
  GPIO.output(24, GPIO.LOW)
  GPIO.output(25, GPIO.LOW)
  GPIO.output(4, GPIO.LOW)
  GPIO.output(7, GPIO.LOW)
  GPIO.output(8, GPIO.LOW)
  GPIO.output(9, GPIO.LOW)
  GPIO.output(10, GPIO.LOW)
  GPIO.output(11, GPIO.LOW)
  GPIO.output(0, GPIO.LOW)
  GPIO.output(1, GPIO.LOW)
  # Light selected LED's
  if count==1 or count==30: GPIO.output(17, GPIO.HIGH)
  if count==2 or count==29: GPIO.output(18, GPIO.HIGH)
  if count==3 or count==28: GPIO.output(21, GPIO.HIGH)
  if count==4 or count==27: GPIO.output(22, GPIO.HIGH)
  if count==5 or count==26: GPIO.output(23, GPIO.HIGH)
  if count==6 or count==25: GPIO.output(24, GPIO.HIGH)
  if count==7 or count==24: GPIO.output(25, GPIO.HIGH)
  if count==8 or count==23: GPIO.output(4, GPIO.HIGH)
  if count==9 or count==22: GPIO.output(7, GPIO.HIGH)
  if count==10 or count==21: GPIO.output(8, GPIO.HIGH)
  if count==11 or count==20: GPIO.output(9, GPIO.HIGH)
  if count==12 or count==19: GPIO.output(10, GPIO.HIGH)
  if count==13 or count==18: GPIO.output(11, GPIO.HIGH)
  if count==14 or count==17: GPIO.output(0, GPIO.HIGH)
  if count==15 or count==16: GPIO.output(1, GPIO.HIGH)
  # Delay
  time.sleep(.5)
```

You can add start and stop switches to make the scanning of the LED's start and stop on your command. The delay time could be reduced to make a wheel of prizes or perhaps even some sort of electronic dice.

I added the switches to the "CE0 and CE1" pins 24 and 26. Here is the schematic diagram.

```
                       470
        Gnd  ──────────/\/\/──────┐
      Pin 11 ──────▷│──────────┤
      Pin 12 ──────▷│──────────┤
      Pin 13 ──────▷│──────────┤
      Pin 15 ──────▷│──────────┤
      Pin 16 ──────▷│──────────┤
      Pin 18 ──────▷│──────────┤
      Pin 22 ──────▷│──────────┤
       Pin 7 ──────▷│──────────┤
                                │
      Pin 26 ─o     o───────────┤
                                │
      Pin 24 ─o     o───────────┘
```

Here is a picture, this time it is using the "Raspberry Pi GPIO extension board". It makes the wiring easier and it looks a lot neater.

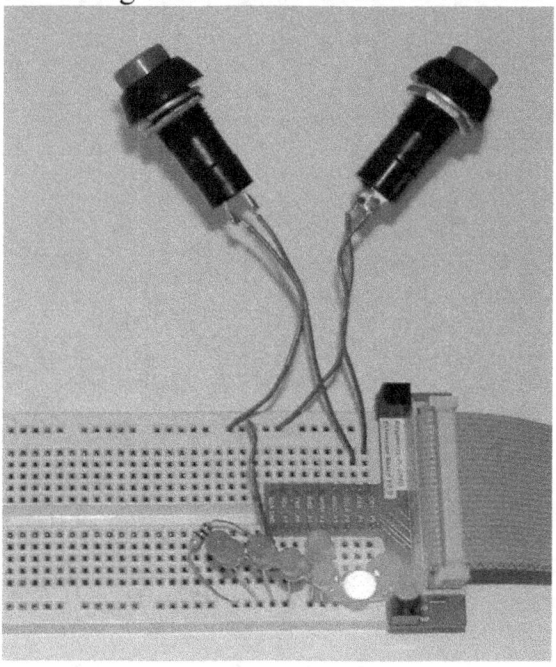

Here is the code to make it work. The switches change the stat of the "run" variable. Then if "run" is enabled the count increases.

```
import RPi.GPIO as GPIO
import time
GPIO.setmode(GPIO.BCM)
GPIO.setup(17, GPIO.OUT)
GPIO.setup(18, GPIO.OUT)
GPIO.setup(27, GPIO.OUT) #21 on older
GPIO.setup(22, GPIO.OUT)
GPIO.setup(23, GPIO.OUT)
GPIO.setup(24, GPIO.OUT)
GPIO.setup(25, GPIO.OUT)
GPIO.setup(4, GPIO.OUT)
# set pins as inputs with pull up resistors
GPIO.setup(7, GPIO.IN, GPIO.PUD_UP)
GPIO.setup(8, GPIO.IN, GPIO.PUD_UP)

# set up loop
run=1
count=1
while count < 10:
  # check the status of the switches
  if GPIO.input(7)==0: run=1
  if GPIO.input(8)==0: run=0
  if run==1: count=count+1
  if count >8: count = 1
  # Turn everything off
  GPIO.output(17, GPIO.LOW)
  GPIO.output(18, GPIO.LOW)
  GPIO.output(27, GPIO.LOW)
  GPIO.output(22, GPIO.LOW)
  GPIO.output(23, GPIO.LOW)
  GPIO.output(24, GPIO.LOW)
  GPIO.output(25, GPIO.LOW)
  GPIO.output(4, GPIO.LOW)
  # Turn on selected LED
  if count==1: GPIO.output(17, GPIO.HIGH)
  if count==2: GPIO.output(18, GPIO.HIGH)
  if count==3: GPIO.output(27, GPIO.HIGH)
  if count==4: GPIO.output(22, GPIO.HIGH)
  if count==5: GPIO.output(23, GPIO.HIGH)
  if count==6: GPIO.output(24, GPIO.HIGH)
  if count==7: GPIO.output(25, GPIO.HIGH)
  if count==8: GPIO.output(4, GPIO.HIGH)
  time.sleep(.1)
# end
```

Chapter 7
LED's with a Shift Register

What if you need to control even more LED's? You can add shift registers to control many more LED's. For this project we will add three shift registers to control a total of 24 LED's.

The 74595 is one of the most popular shift register that is used to run LED arrays. The shift register has serial data input, clock input, and latch inputs. They require 5 volts to operate but they run fine with the 3.3 volt logic coming from the Raspberry Pi. Here is a drawing showing the actual pin connections to the 74595 and below that there is the schematic diagram representation of the IC.

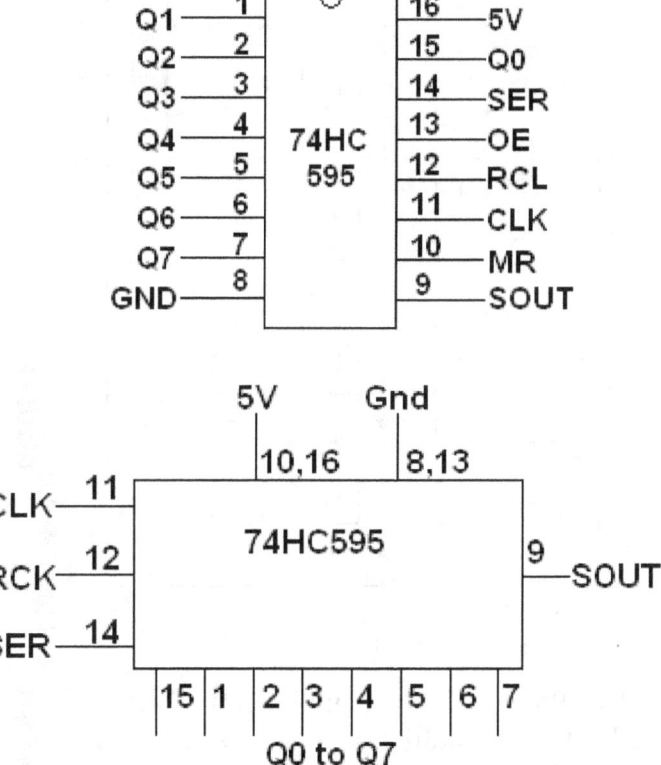

The guts of the 74HC595 are somewhat complicated. The internal schematic of the IC is shown in the next drawing.

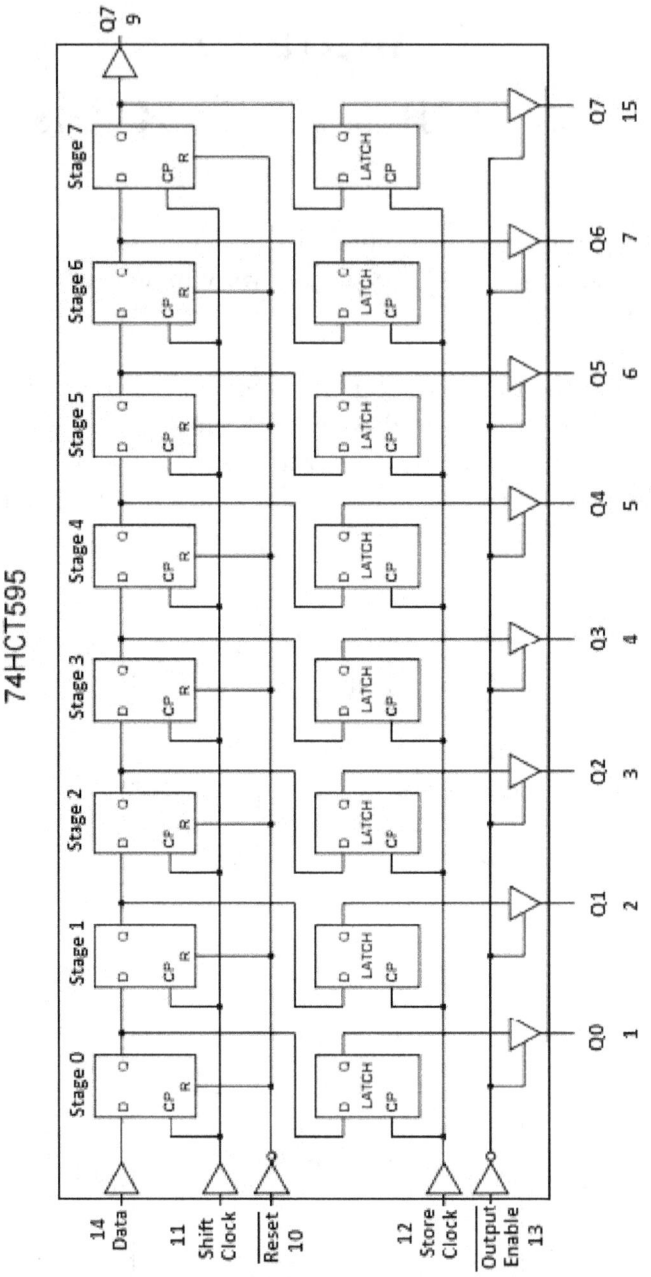

Inside of the 74HC595 besides the eight latches of the shift register there are eight additional latches that output the contents of the shift register. Then there are eight tri-state output drivers. Those latches and drivers are shown in the previous drawing. Having the extra latches in the middle prevents what is being shifted through the shift register from being displayed until it is shifted into place.

The shift registers will be connected in series so the data can be shifted down through them. That is done by connecting pin 9 to pin 14 of the next shift register. Here is the schematic diagram.

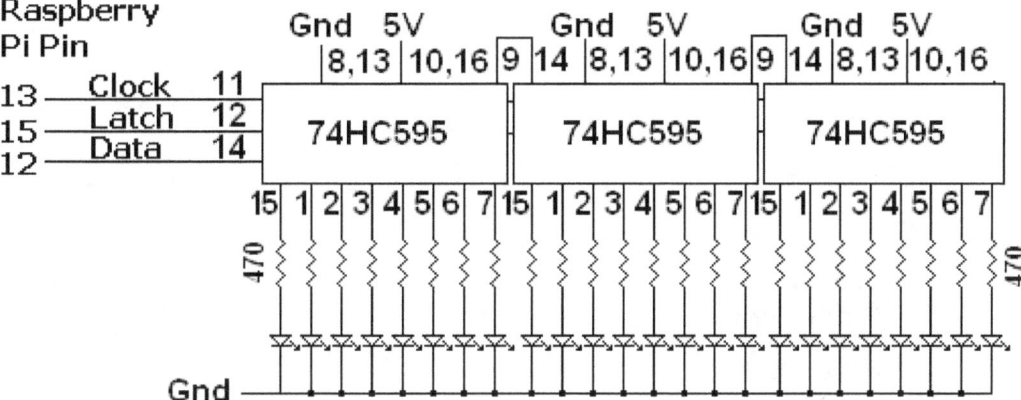

Here is a picture of the LED's with the shift register's in operation.

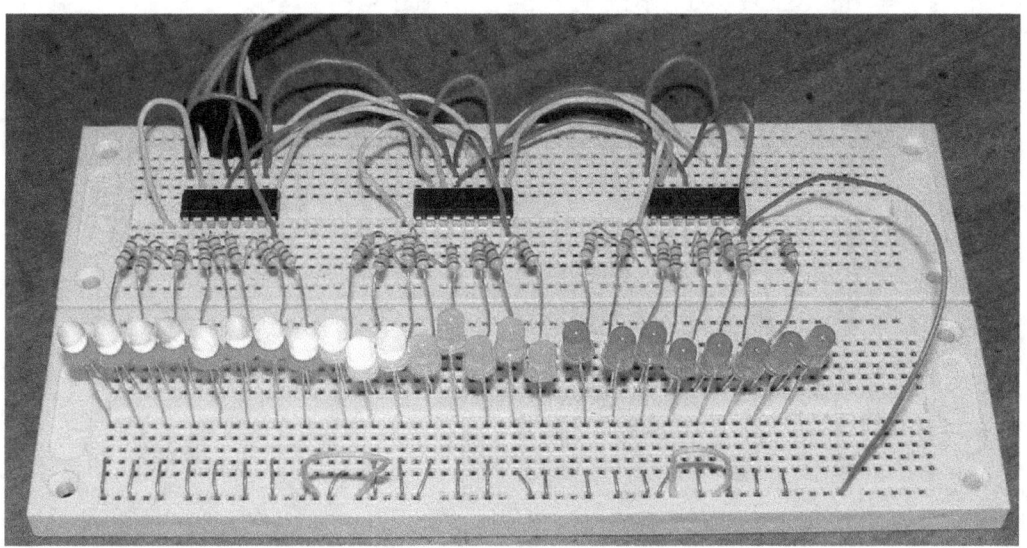

Here is the Python code to make it work. If you replace "if shift <= scan:" with "if shift == scan:" then only one LED will light at a time. As it is coded the LED's light up progressively then the go out in the same order.

```
# LED with 74595
# Uses 74595 Shift Registers
import RPi.GPIO as GPIO
```

```python
import time
GPIO.setmode(GPIO.BCM)

GPIO.setup(17, GPIO.OUT) # Serial Data
GPIO.setup(18, GPIO.OUT) # Not used
GPIO.setup(27, GPIO.OUT) # Clock, 21 on older
GPIO.setup(22, GPIO.OUT) # Latch

# set up the loop
cycle= 0
scan = 1
direction=0

while cycle < 1000:
    if scan==24:
        direction=0
    if scan==0:
        direction=1
    # Send data to the shift registers
    shift = 24
    while shift >= 0:
        GPIO.output(17, GPIO.LOW)
        # determine if bit is set or clear
        if shift <= scan:
            GPIO.output(17, GPIO.HIGH)
        # advance the clock
        GPIO.output(27, GPIO.LOW)
        GPIO.output(27, GPIO.HIGH)
        shift=shift-1
    # Latch and display the output
    GPIO.output(22, GPIO.LOW)
    GPIO.output(22, GPIO.HIGH)
    if direction==0:scan=scan-1
    if direction==1:scan=scan+1
    time.sleep(.05)
    cycle=cycle+1
# end
```

Chapter 8
Seven Segment LED Arrays

For this chapter we are going to add some 7 segment LED arrays. These are basically 7 or 8 LED's arranged so that when the right LED's are lit up they form the numbers 0 to 9. These devices make it much easier to display numbers. They were first used in calculators, but now are found in lots of places.

Here is the pinout diagram for the SMA42056 Seven segment LED arrays that can be used for this project. You can use any seven segment array that you want you just need to figure out how to connect to it. A 9 volt battery and a 470 ohm resistor is the best method to figure out what pin goes where. The orientation can be set up by the decimal point being in the lower right corner.

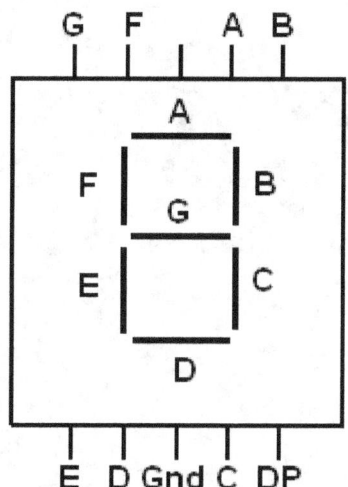

Up next is the schematic diagram of the 3 digit counter.

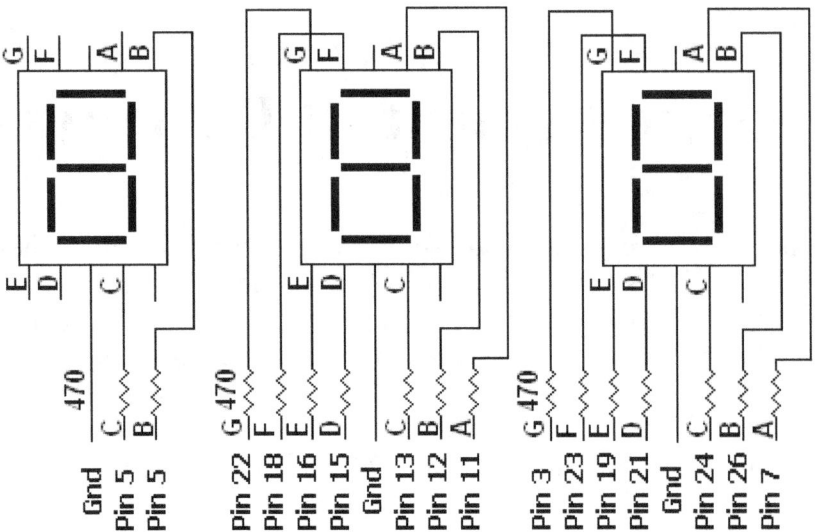

Here is what the assembled 3 digit counter hardware looks like.

This Python program will count from 1 to 199

```
# 7seg.py – Seven Segment
# Counts to 199 using the GPIO pins.
```

```python
import RPi.GPIO as GPIO
import time
GPIO.setmode(GPIO.BCM)

# Leading digit 0 or 1
GPIO.setup(1, GPIO.OUT)

# Second digit A-F
GPIO.setmode(GPIO.BCM)
GPIO.setup(17, GPIO.OUT)
GPIO.setup(18, GPIO.OUT)
GPIO.setup(21, GPIO.OUT)
GPIO.setup(22, GPIO.OUT)
GPIO.setup(23, GPIO.OUT)
GPIO.setup(24, GPIO.OUT)
GPIO.setup(25, GPIO.OUT)

# Third digit A-F
GPIO.setup(4, GPIO.OUT)
GPIO.setup(7, GPIO.OUT)
GPIO.setup(8, GPIO.OUT)
GPIO.setup(9, GPIO.OUT)
GPIO.setup(10, GPIO.OUT)
GPIO.setup(11, GPIO.OUT)
GPIO.setup(0, GPIO.OUT)

# set up loop
count=0
while count < 200:
  # turn everything off
  GPIO.output(1, GPIO.LOW)
  GPIO.output(0, GPIO.LOW)
  GPIO.output(17, GPIO.LOW)
  GPIO.output(18, GPIO.LOW)
  GPIO.output(21, GPIO.LOW)
  GPIO.output(22, GPIO.LOW)
  GPIO.output(23, GPIO.LOW)
  GPIO.output(24, GPIO.LOW)
  GPIO.output(25, GPIO.LOW)
  GPIO.output(4, GPIO.LOW)
  GPIO.output(7, GPIO.LOW)
  GPIO.output(8, GPIO.LOW)
  GPIO.output(9, GPIO.LOW)
  GPIO.output(10, GPIO.LOW)
  GPIO.output(11, GPIO.LOW)

  # turn on segments
  # First digit
  if count >= 100:
    GPIO.output(1, GPIO.HIGH)
```

```python
# second digit divide by 10 then get right digit
ccount=count/10%10
if ccount==0:
  GPIO.output(17, GPIO.HIGH)
  GPIO.output(18, GPIO.HIGH)
  GPIO.output(21, GPIO.HIGH)
  GPIO.output(22, GPIO.HIGH)
  GPIO.output(23, GPIO.HIGH)
  GPIO.output(24, GPIO.HIGH)
if ccount==1:
  GPIO.output(18, GPIO.HIGH)
  GPIO.output(21, GPIO.HIGH)
if ccount==2:
  GPIO.output(17, GPIO.HIGH)
  GPIO.output(18, GPIO.HIGH)
  GPIO.output(22, GPIO.HIGH)
  GPIO.output(23, GPIO.HIGH)
  GPIO.output(25, GPIO.HIGH)
if ccount==3:
  GPIO.output(17, GPIO.HIGH)
  GPIO.output(18, GPIO.HIGH)
  GPIO.output(22, GPIO.HIGH)
  GPIO.output(21, GPIO.HIGH)
  GPIO.output(25, GPIO.HIGH)
if ccount==4:
  GPIO.output(18, GPIO.HIGH)
  GPIO.output(21, GPIO.HIGH)
  GPIO.output(24, GPIO.HIGH)
  GPIO.output(25, GPIO.HIGH)
if ccount==5:
  GPIO.output(17, GPIO.HIGH)
  GPIO.output(21, GPIO.HIGH)
  GPIO.output(22, GPIO.HIGH)
  GPIO.output(24, GPIO.HIGH)
  GPIO.output(25, GPIO.HIGH)
if ccount==6:
  GPIO.output(17, GPIO.HIGH)
  GPIO.output(21, GPIO.HIGH)
  GPIO.output(22, GPIO.HIGH)
  GPIO.output(23, GPIO.HIGH)
  GPIO.output(24, GPIO.HIGH)
  GPIO.output(25, GPIO.HIGH)
if ccount==7:
  GPIO.output(17, GPIO.HIGH)
  GPIO.output(18, GPIO.HIGH)
  GPIO.output(21, GPIO.HIGH)
if ccount==8:
  GPIO.output(17, GPIO.HIGH)
  GPIO.output(18, GPIO.HIGH)
```

```python
    GPIO.output(21, GPIO.HIGH)
    GPIO.output(22, GPIO.HIGH)
    GPIO.output(23, GPIO.HIGH)
    GPIO.output(24, GPIO.HIGH)
    GPIO.output(25, GPIO.HIGH)
if ccount==9:
    GPIO.output(17, GPIO.HIGH)
    GPIO.output(18, GPIO.HIGH)
    GPIO.output(21, GPIO.HIGH)
    GPIO.output(24, GPIO.HIGH)
    GPIO.output(25, GPIO.HIGH)

# third digit get right digit
ccount=count%10
if ccount==0:
    GPIO.output(4, GPIO.HIGH)
    GPIO.output(7, GPIO.HIGH)
    GPIO.output(8, GPIO.HIGH)
    GPIO.output(9, GPIO.HIGH)
    GPIO.output(10, GPIO.HIGH)
    GPIO.output(11, GPIO.HIGH)
if ccount==1:
    GPIO.output(7, GPIO.HIGH)
    GPIO.output(8, GPIO.HIGH)
if ccount==2:
    GPIO.output(4, GPIO.HIGH)
    GPIO.output(7, GPIO.HIGH)
    GPIO.output(9, GPIO.HIGH)
    GPIO.output(10, GPIO.HIGH)
    GPIO.output(0, GPIO.HIGH)
if ccount==3:
    GPIO.output(4, GPIO.HIGH)
    GPIO.output(7, GPIO.HIGH)
    GPIO.output(9, GPIO.HIGH)
    GPIO.output(8, GPIO.HIGH)
    GPIO.output(0, GPIO.HIGH)
if ccount==4:
    GPIO.output(7, GPIO.HIGH)
    GPIO.output(8, GPIO.HIGH)
    GPIO.output(11, GPIO.HIGH)
    GPIO.output(0, GPIO.HIGH)
if ccount==5:
    GPIO.output(4, GPIO.HIGH)
    GPIO.output(8, GPIO.HIGH)
    GPIO.output(9, GPIO.HIGH)
    GPIO.output(11, GPIO.HIGH)
    GPIO.output(0, GPIO.HIGH)
if ccount==6:
    GPIO.output(4, GPIO.HIGH)
    GPIO.output(8, GPIO.HIGH)
```

```
      GPIO.output(9, GPIO.HIGH)
      GPIO.output(10, GPIO.HIGH)
      GPIO.output(11, GPIO.HIGH)
      GPIO.output(0, GPIO.HIGH)
    if ccount==7:
      GPIO.output(4, GPIO.HIGH)
      GPIO.output(7, GPIO.HIGH)
      GPIO.output(8, GPIO.HIGH)
    if ccount==8:
      GPIO.output(4, GPIO.HIGH)
      GPIO.output(7, GPIO.HIGH)
      GPIO.output(8, GPIO.HIGH)
      GPIO.output(9, GPIO.HIGH)
      GPIO.output(10, GPIO.HIGH)
      GPIO.output(11, GPIO.HIGH)
      GPIO.output(0, GPIO.HIGH)
    if ccount==9:
      GPIO.output(4, GPIO.HIGH)
      GPIO.output(7, GPIO.HIGH)
      GPIO.output(8, GPIO.HIGH)
      GPIO.output(11, GPIO.HIGH)
      GPIO.output(0, GPIO.HIGH)
    # delay
    time.sleep(1)
    count=count+1
# end
```

Chapter 9
Seven Segment Arrays
With Shift Registers

What if you want more than two LED arrays? We can add some 74595 shift registers and run as many LED arrays as we want to. This project will run six shift registers and hence six 7 segment LED arrays. The code will display the time on the LED arrays.

Here is a picture of the six 7 segment display in operation. The LED arrays are in front because there are 4 segments connected on their back side and there are three on their front side. So there are three spaces between the sets of displays to enable routing the signals around from the shift registers.

Up next is the schematic diagram. It is turned sideways because the schematic is rather large.

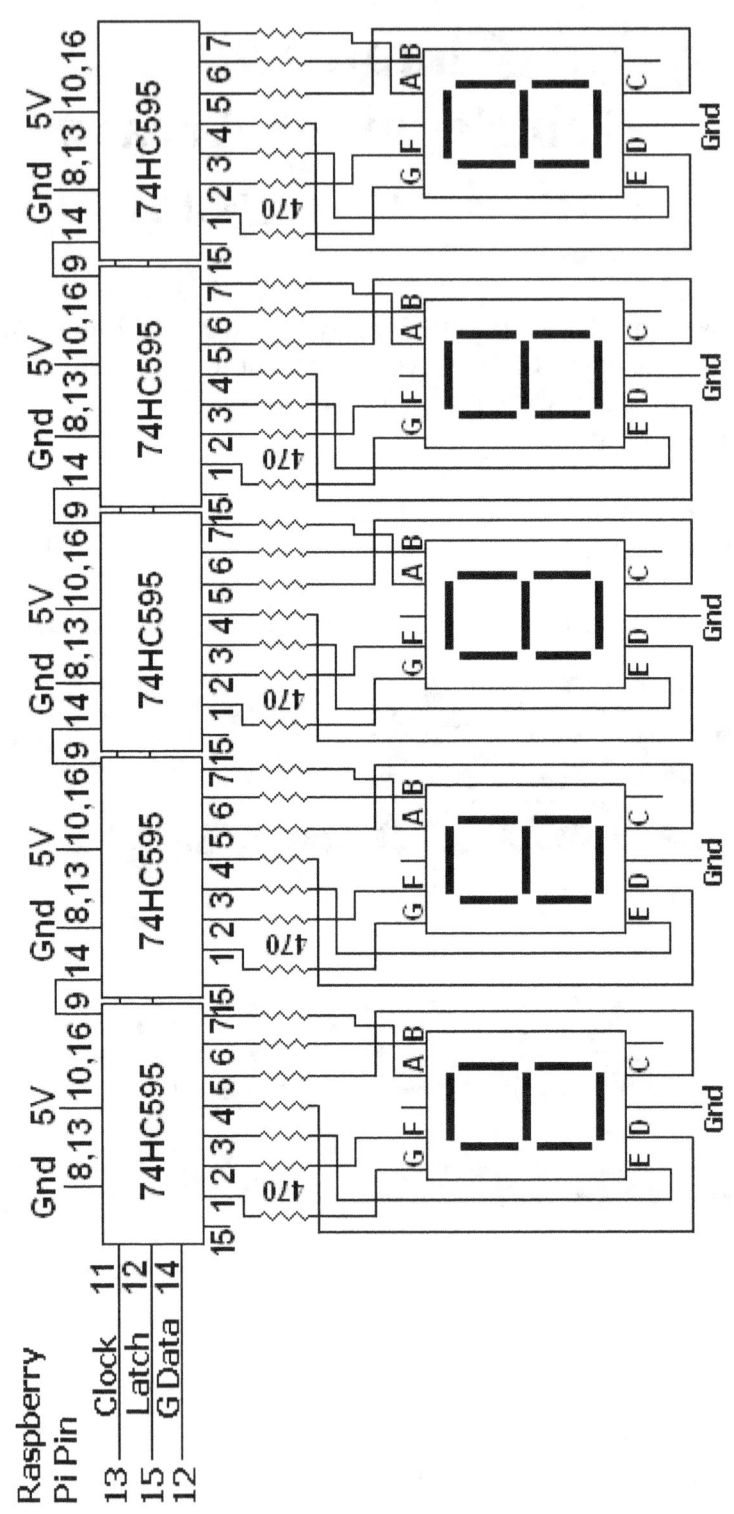

Here is the code to make it work. Note that is it much shorter that some other programs. That is from the use of a "dictionaries" in the "Nums" array. Each number value has a matching string. This makes the software listing much shorter.

```
# LED7segTime
# Uses segment strings strung together
import RPi.GPIO as GPIO
import time
GPIO.setmode(GPIO.BCM)

GPIO.setup(17, GPIO.OUT) # Serial Data
GPIO.setup(18, GPIO.OUT) #
GPIO.setup(27, GPIO.OUT) # Clock, 21 on older
GPIO.setup(22, GPIO.OUT) # Latch

# Segments 1 nul then: G-F-E-D-C-B-A
Nums = {'0':(0,0,1,1,1,1,1,1),
        '1':(0,0,0,0,0,1,1,0),
        '2':(0,1,0,1,1,0,1,1),
        '3':(0,1,0,0,1,1,1,1),
        '4':(0,1,1,0,0,1,1,0),
        '5':(0,1,1,0,1,1,0,1),
        '6':(0,1,1,1,1,1,0,1),
        '7':(0,0,0,0,0,1,1,1),
        '8':(0,1,1,1,1,1,1,1),
        '9':(0,1,1,0,0,1,1,1)}

print 'the current time is: '
print time.strftime( '%H:%M:%S' )

# set up the loop
while True:
    tstr=time.strftime( '%H:%M:%S')
    Data1=Nums[tstr[0]]+ Nums[tstr[1]]+ Nums[tstr[3]]+
Nums[tstr[4]]+ Nums[tstr[6]]+ Nums[tstr[7]]
    # Send data to the shift registers
    shift = 47
    while shift >= 0:
        GPIO.output(17, GPIO.LOW)
        # determine if bit is set or clear
        if Data1[shift] == 1: GPIO.output(17, GPIO.HIGH)
        # advance the clock
        GPIO.output(27, GPIO.LOW); GPIO.output(27, GPIO.HIGH)
        shift=shift-1
    # Latch and display the output
    GPIO.output(22, GPIO.LOW); GPIO.output(22, GPIO.HIGH)
    time.sleep(.1)
# end
```

Chapter 10
Seven Segment Arrays
Multiplexed

Another method of running multiple seven segment arrays is to multiplex them. Basically what you do is to display your data on one array, then on the next array then on the next array so fast that they all appear to be on. This reduces the pins needed because you only need 7 pins for the seven segments and then a select pin for each array.

In this experiment we are going to do two things. First we will get an analog reading using the MCP3008 and then we will display the results on three seven segment LED arrays. For the analog section schematic see the chapter on adding analog ports.

We will need to introduce a new IC, it is the ULN2003. It is equivalent to several power transistors inside of one case. It can sink 1/2 an amp per output. This is ideal for running multiple LED's as the Raspberry Pi can only run one LED per output. The ULN2003 schematic looks like this.

ULN2003

Here is the schematic diagram of the multiplexed seven segment array section of this project.

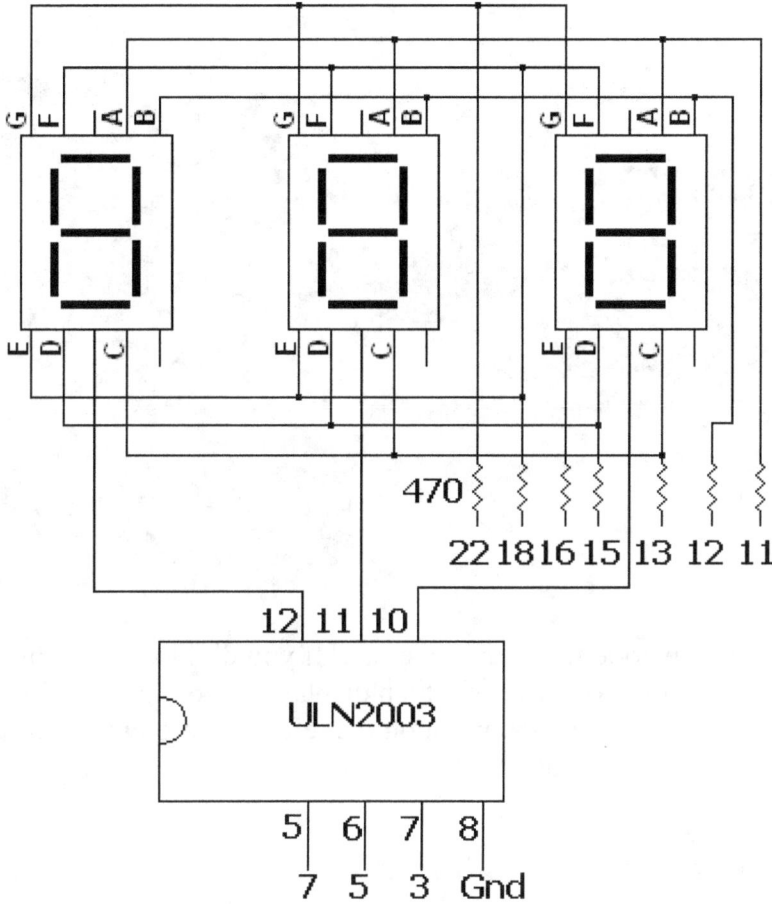

Here is a picture of the multiplexed arrays and the analog input wired up and working. The ULN2003 is on the left, the three seven segment arrays are in the left middle and the MCP3008 is on the right middle. I used the GPIO extension to make the wiring easier to do and to look neater. A variable resistor was used to vary the output from 0 to 3.3 volts.

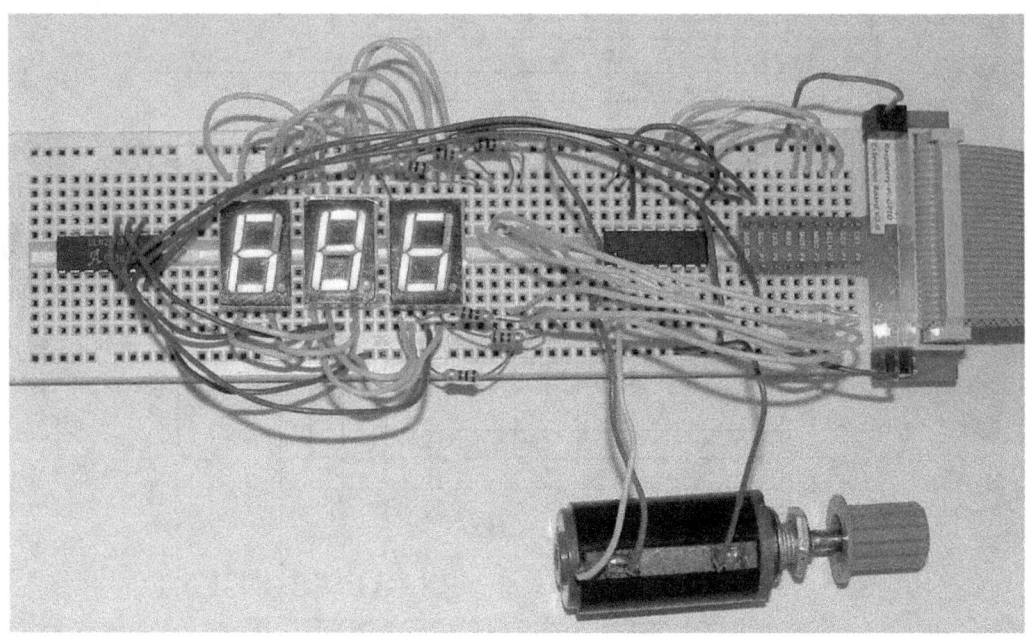

Here is the python code to make it operate. If you divide the results from the analog to digital converter you will then display the voltage in volts. You could even jumper the decimal point on the left digit through a resistor to 3.3 volts so that it will say "3.30" volts.

```
# 7 Segment Analog
# Read analog port 0 and display the results

import RPi.GPIO as GPIO
import time
GPIO.setmode(GPIO.BCM)

# Segments A-F
GPIO.setup(17, GPIO.OUT)
GPIO.setup(18, GPIO.OUT)
GPIO.setup(27, GPIO.OUT) #21 on older
GPIO.setup(22, GPIO.OUT)
GPIO.setup(23, GPIO.OUT)
GPIO.setup(24, GPIO.OUT)
GPIO.setup(25, GPIO.OUT)

# digit selection pins
GPIO.setup(4, GPIO.OUT)
GPIO.setup(2, GPIO.OUT)
GPIO.setup(3, GPIO.OUT)

# Analog pins
GPIO.setup(11, GPIO.OUT) # Sclock
```

```
GPIO.setup(9, GPIO.IN) # MISO
GPIO.setup(10, GPIO.OUT) # MOSI
GPIO.setup(8, GPIO.OUT) # CE0
port=0
serout=0x18

# set up loop
cycle=0
while cycle < 10:
  # only sample every 10 times through
  cycle=cycle+1
  if cycle == 10: cycle = 0
  if cycle == 1:
    # Read analog data
    GPIO.output(8, GPIO.HIGH) # deselect chip
    GPIO.output(11, GPIO.LOW) # set clock low
    shift=0
    adcin=0
    # 24 pits shifted in and out
    while shift < 24:
      GPIO.output(8, GPIO.LOW) # select chip
      # low for most bits out
      GPIO.output(10, GPIO.LOW)
      if (shift == 7 or shift ==8):
        GPIO.output(10, GPIO.HIGH)
      if shift == 9:
        if (port > 3):
          GPIO.output(10, GPIO.HIGH)
      if shift == 10:
        if (port & 0x02):
          GPIO.output(10, GPIO.HIGH)
      if shift == 11:
        if (port & 0x01):
          GPIO.output(10, GPIO.HIGH)
      if shift > 13: # last 10 bits
        if(GPIO.input(9)):
          adcin = adcin+1 #set bit
        adcin = adcin << 1 # left shift 1
      # cycle the clock
      GPIO.output(11, GPIO.LOW)
      GPIO.output(11, GPIO.HIGH)
      shift=shift+1
    print adcin
  digit=1

# now display the results
  while digit < 4:
    # turn everything off
    GPIO.output(17, GPIO.LOW)
    GPIO.output(18, GPIO.LOW)
```

```python
      GPIO.output(27, GPIO.LOW)
      GPIO.output(22, GPIO.LOW)
      GPIO.output(23, GPIO.LOW)
      GPIO.output(24, GPIO.LOW)
      GPIO.output(25, GPIO.LOW)
      GPIO.output(4, GPIO.LOW)
      GPIO.output(2, GPIO.LOW)
      GPIO.output(3, GPIO.LOW)

    if digit == 1:
      # first digit get right most digit
      ccount=adcin%10
      GPIO.output(2, GPIO.HIGH)
    if digit == 2:
      # second digit divide by 10 then get digit
      ccount=adcin/10%10
      GPIO.output(3, GPIO.HIGH)
    if digit == 3:
      # third digit divide by 100 then get digit
      ccount=adcin/100%10
      GPIO.output(4, GPIO.HIGH)
    # turn on segments
    if ccount==0:
      GPIO.output(17, GPIO.HIGH);GPIO.output(18,
GPIO.HIGH);GPIO.output(27, GPIO.HIGH)
      GPIO.output(22, GPIO.HIGH);GPIO.output(23,
GPIO.HIGH);GPIO.output(24, GPIO.HIGH)
    if ccount==1:
      GPIO.output(18, GPIO.HIGH);GPIO.output(27, GPIO.HIGH)
    if ccount==2:
      GPIO.output(17, GPIO.HIGH);GPIO.output(18,
GPIO.HIGH);GPIO.output(22, GPIO.HIGH)
      GPIO.output(23, GPIO.HIGH);GPIO.output(25, GPIO.HIGH)
    if ccount==3:
      GPIO.output(17, GPIO.HIGH);GPIO.output(18,
GPIO.HIGH);GPIO.output(22, GPIO.HIGH)
      GPIO.output(27, GPIO.HIGH);GPIO.output(25, GPIO.HIGH)
    if ccount==4:
      GPIO.output(18, GPIO.HIGH);GPIO.output(27, GPIO.HIGH);
      GPIO.output(24, GPIO.HIGH);GPIO.output(25, GPIO.HIGH)
    if ccount==5:
      GPIO.output(17, GPIO.HIGH);GPIO.output(27,
GPIO.HIGH);GPIO.output(22, GPIO.HIGH)
      GPIO.output(24, GPIO.HIGH);GPIO.output(25, GPIO.HIGH);
    if ccount==6:
      GPIO.output(17, GPIO.HIGH);GPIO.output(27,
GPIO.HIGH);GPIO.output(22, GPIO.HIGH)
      GPIO.output(23, GPIO.HIGH);GPIO.output(24,
GPIO.HIGH);GPIO.output(25, GPIO.HIGH)
    if ccount==7:
```

```python
        GPIO.output(17, GPIO.HIGH);GPIO.output(18,
GPIO.HIGH);GPIO.output(27, GPIO.HIGH)
    if ccount==8:
        GPIO.output(17, GPIO.HIGH);GPIO.output(18,
GPIO.HIGH);GPIO.output(27, GPIO.HIGH)
        GPIO.output(22, GPIO.HIGH);GPIO.output(23, GPIO.HIGH)
        GPIO.output(24, GPIO.HIGH);GPIO.output(25, GPIO.HIGH)
    if ccount==9:
        GPIO.output(17, GPIO.HIGH);GPIO.output(18,
GPIO.HIGH);GPIO.output(27, GPIO.HIGH);
        GPIO.output(24, GPIO.HIGH);GPIO.output(25, GPIO.HIGH)
    digit=digit+1
    time.sleep(.005)
# end
```

Chapter 11
5x7 LED Array

LED matrix arrays are made out of several LED's arranged into columns and rows. Since each LED cannot be accessed individually they must be accessed by the rows. First one row is lit up then the next row is lit. This is done fast enough that the rows appear to be continuously lit.

It is possible to directly drive a LED array with a Raspberry Pi. To be honest the Raspberry Pi's 3.3 volts and low driving current create issues. We have been using 470 ohm resistors for 5 volt operation. The correct value for 3.3 volt operation is 170 ohms. However since each LED is only on 1/7 of the time, we will reduce that to 100 ohms. We could use 25 ohms as 160/7 is 23. I tested using 25 ohm resistors and they do work. However the Raspberry Pi will not provide the needed current to make it very bright.

The 5x7 LED array that we are using is a LTP2157AG. However any LED array would work you just need to figure out the pinout. Observe that two pins come out in two different locations (R4 and C3). You only need to connect to one of each of them.

Here are the pin connections arranged as they actually appear from above for the LTP-2157. Pin one is the pin closest to the identification markings.

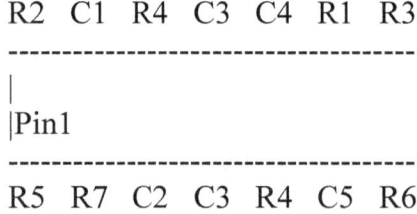

```
R2  C1  R4  C3  C4  R1  R3
--------------------------------------
|
|Pin1
--------------------------------------
R5  R7  C2  C3  R4  C5  R6
```

Up next is the schematic diagram of the LTP-2157

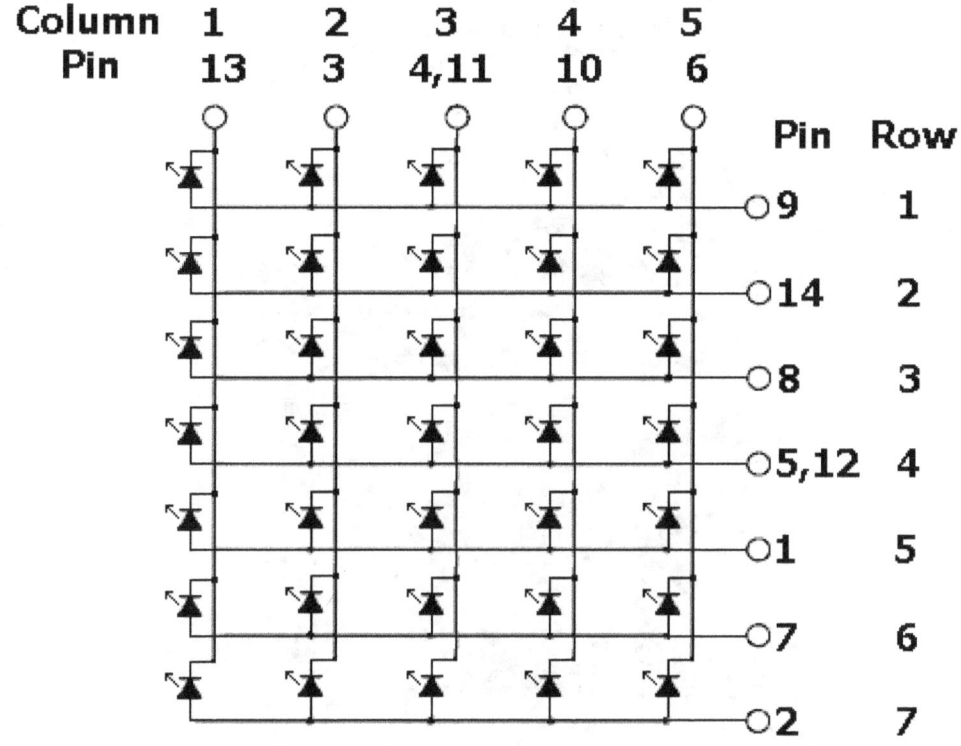

This is how to hook it up. Do not forget to use 100 ohm resistors in the column connections!

Array	Pin	GPIO
R1	11	17
R2	12	18
R3	13	21/27
R4	15	22
R5	16	23
R6	18	24
R7	22	25
C1	7	4
C2	19	10
C3	21	9
C4	23	11
C5	8	8

Here is a picture of it in operation. There are resistors in the column circuits and bare jumper wires in the row circuits.

Here is the code to make it work. Note that the code is a bit long, we will be adding arrays in future projects that will make the code much shorter. This code produces the letter "R"

```
# LED5x7 Array demonstration
import RPi.GPIO as GPIO
import time
# rows lit when high
GPIO.setmode(GPIO.BCM)
GPIO.setup(17, GPIO.OUT)
GPIO.setup(18, GPIO.OUT)
GPIO.setup(27, GPIO.OUT)#21 on older
GPIO.setup(22, GPIO.OUT)
GPIO.setup(23, GPIO.OUT)
GPIO.setup(24, GPIO.OUT)
GPIO.setup(25, GPIO.OUT)
#columns lit when low
GPIO.setup(4, GPIO.OUT)
GPIO.setup(10, GPIO.OUT)
GPIO.setup(9, GPIO.OUT)
GPIO.setup(11, GPIO.OUT)
GPIO.setup(8, GPIO.OUT)
```

```
while True:
    GPIO.output(4, GPIO.LOW)
    GPIO.output(10, GPIO.LOW)
    GPIO.output(9, GPIO.LOW)
    GPIO.output(11, GPIO.LOW)
    GPIO.output(8, GPIO.HIGH)
# turn on row1
    GPIO.output(25, GPIO.HIGH)
    time.sleep(.001)
    GPIO.output(25, GPIO.LOW)
# next columns
    GPIO.output(4, GPIO.LOW)
    GPIO.output(10, GPIO.HIGH)
    GPIO.output(9, GPIO.HIGH)
    GPIO.output(11, GPIO.HIGH)
    GPIO.output(8, GPIO.LOW)
# turn on row2
    GPIO.output(24, GPIO.HIGH)
    time.sleep(.001)
    GPIO.output(24, GPIO.LOW)
# next columns
    GPIO.output(4, GPIO.LOW)
    GPIO.output(10, GPIO.HIGH)
    GPIO.output(9, GPIO.HIGH)
    GPIO.output(11, GPIO.HIGH)
    GPIO.output(8, GPIO.LOW)
# turn on row3
    GPIO.output(23, GPIO.HIGH)
    time.sleep(.001)
    GPIO.output(23, GPIO.LOW)
# next columns
    GPIO.output(4, GPIO.LOW)
    GPIO.output(10, GPIO.LOW)
    GPIO.output(9, GPIO.LOW)
    GPIO.output(11, GPIO.LOW)
    GPIO.output(8, GPIO.HIGH)
# turn on row4
    GPIO.output(22, GPIO.HIGH)
    time.sleep(.001)
    GPIO.output(22, GPIO.LOW)
# next columns
    GPIO.output(4, GPIO.LOW)
    GPIO.output(10, GPIO.HIGH)
    GPIO.output(9, GPIO.HIGH)
    GPIO.output(11, GPIO.HIGH)
    GPIO.output(8, GPIO.LOW)
# turn on row5
    GPIO.output(27, GPIO.HIGH)
    time.sleep(.001)
```

```
    GPIO.output(27, GPIO.LOW)
# next columns
    GPIO.output(4, GPIO.LOW)
    GPIO.output(10, GPIO.HIGH)
    GPIO.output(9, GPIO.HIGH)
    GPIO.output(11, GPIO.HIGH)
    GPIO.output(8, GPIO.LOW)
# turn on row6
    GPIO.output(18, GPIO.HIGH)
    time.sleep(.001)
    GPIO.output(18, GPIO.LOW)
# next columns
    GPIO.output(4, GPIO.LOW)
    GPIO.output(10, GPIO.HIGH)
    GPIO.output(9, GPIO.HIGH)
    GPIO.output(11, GPIO.HIGH)
    GPIO.output(8, GPIO.LOW)
# turn on row7
    GPIO.output(17, GPIO.HIGH)
    time.sleep(.001)
    GPIO.output(17, GPIO.LOW)
# end
```

Chapter 12
5x7 Array with a Shift Register

What will happen if we add a shift register to drive the 5x7 LED array? It will help to make it a lot brighter. That is because the shift register will provide 5 volt logic levels to the rows.

This is the schematic diagram. Don't forget to add the 100 ohm resistors in the columns, they are the first 5 connections shown.

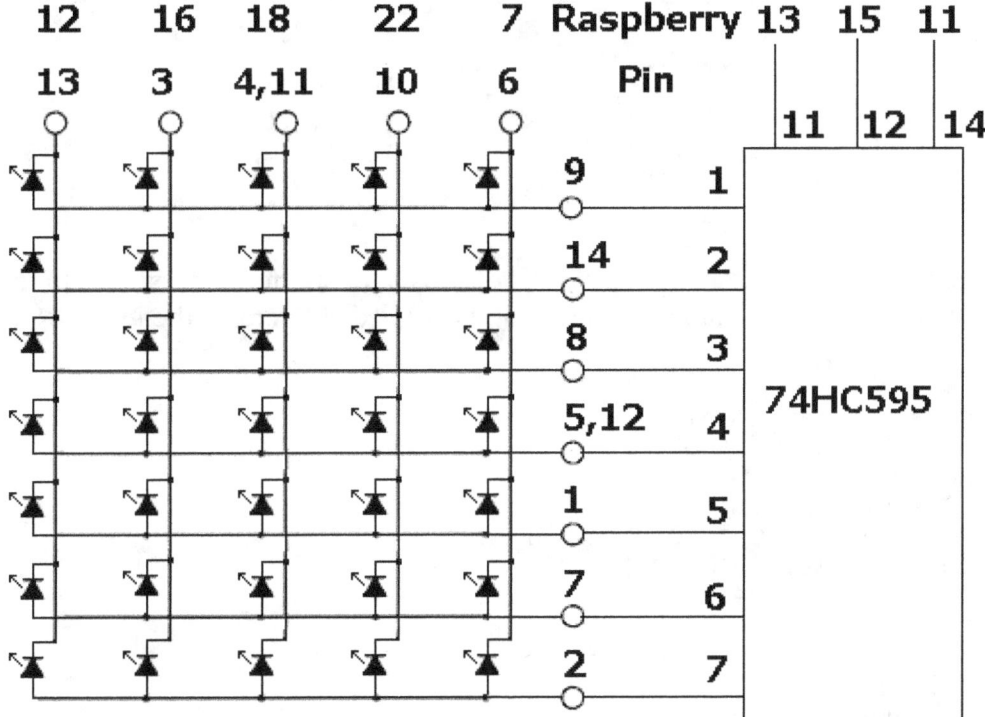

Up next is a picture of the 5x7 LED array with the shift register located in the upper right corner.

Here is the code to make it work. The code is a lot shorter because this time I used an array that contains all the 1's and 0's needed to create the letter "R".

```
# LED 5x7 Array with 74595
# Uses one 74595 Shift Register
import RPi.GPIO as GPIO
import time
GPIO.setmode(GPIO.BCM)
GPIO.setup(17, GPIO.OUT) # Serial Data
GPIO.setup(18, GPIO.OUT) # C1
GPIO.setup(27, GPIO.OUT) # Clock, 21 on older
GPIO.setup(22, GPIO.OUT) # Latch
GPIO.setup(23, GPIO.OUT) # C2
GPIO.setup(24, GPIO.OUT) # C3
GPIO.setup(25, GPIO.OUT) # C4
GPIO.setup(4, GPIO.OUT) # C5

R1 =[1,1,1,1,0,
     1,0,0,0,1,
     1,0,0,0,1,
     1,1,1,1,0,
     1,0,0,0,1,
     1,0,0,0,1,
     1,0,0,0,1]

# set up the loop
```

```
cycle= 0
while cycle < 10000:
  row = 0
  while row < 7:
    # determine if bit is set or clear low=lit
    GPIO.output(18, GPIO.HIGH)
    GPIO.output(23, GPIO.HIGH)
    GPIO.output(24, GPIO.HIGH)
    GPIO.output(25, GPIO.HIGH)
    GPIO.output(4, GPIO.HIGH)
    if R1[row*5] == 1: GPIO.output(18, GPIO.LOW)
    if R1[1+row*5] == 1: GPIO.output(23, GPIO.LOW)
    if R1[2+row*5] == 1: GPIO.output(24, GPIO.LOW)
    if R1[3+row*5] == 1: GPIO.output(25, GPIO.LOW)
    if R1[4+row*5] == 1: GPIO.output(4, GPIO.LOW)
    # Send data to the shift register only for row 0
    GPIO.output(17, GPIO.LOW)
    if row==0: GPIO.output(17, GPIO.HIGH)
    # Advance the clock
    GPIO.output(27, GPIO.LOW)
    GPIO.output(27, GPIO.HIGH)
    # latch and display the data
    GPIO.output(22, GPIO.LOW)
    GPIO.output(22, GPIO.HIGH)
    time.sleep(.001)
    row = row+1
  cycle=cycle+1
# end
```

Chapter 13
8x16 Dual Color LED Array

LED arrays have columns and rows with LED's at the cross points. The Raspberry Pi would quickly run out of pins when driving these LED arrays so we will use shift registers to interface to the LED arrays. This project can start with just one 5x8 LED array then be expanded to the full 8x15 using three 5x8 dual color LED arrays.

We will be using three CSM-58261's or 172054's dual color 5 by 8 LED arrays. They are arranged to be 15 by 8 in total size. These dual color LED arrays are getting to be hard to find but this design will work with any common anode dual color LED array. Just use a 9 volt battery and a 470 ohm resistor to figure out the pinout if you do not know what it is.

Here is the pin out for one of the LED arrays. The rows are called "D1-D8" so they will not be confused with the Red columns

R1 D3 D1 D2 D4 G3 G4 R5 G5

         O   O   O   O   O
         O   O   O   O   O
         O   O   O   O   O
         O   O   O   O   O
         O   O   O   O   O
         O   O   O   O   O
         O   O   O   O   O
         O   O   O   O   O

G1 G2 R2 R3 D6 D7 R4 D5 D8

When three of them are lined up the pins will need to be redefined like what this next drawing shows.

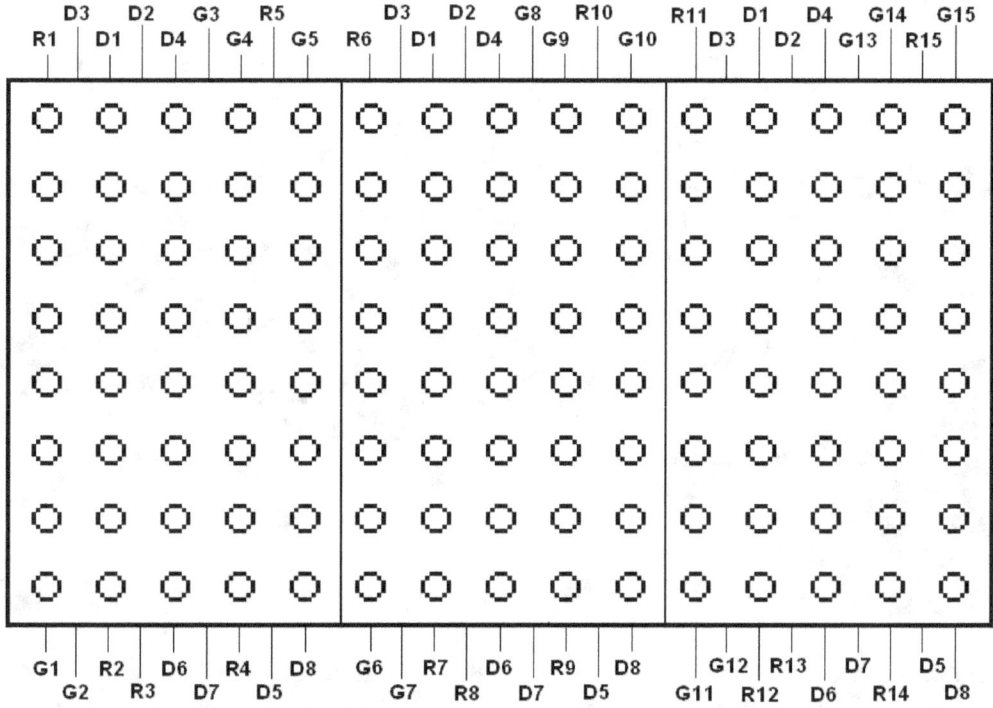

Here is a picture of the completed project. The row selecting 74595 is located on a separate breadboard off to the left side of the picture.

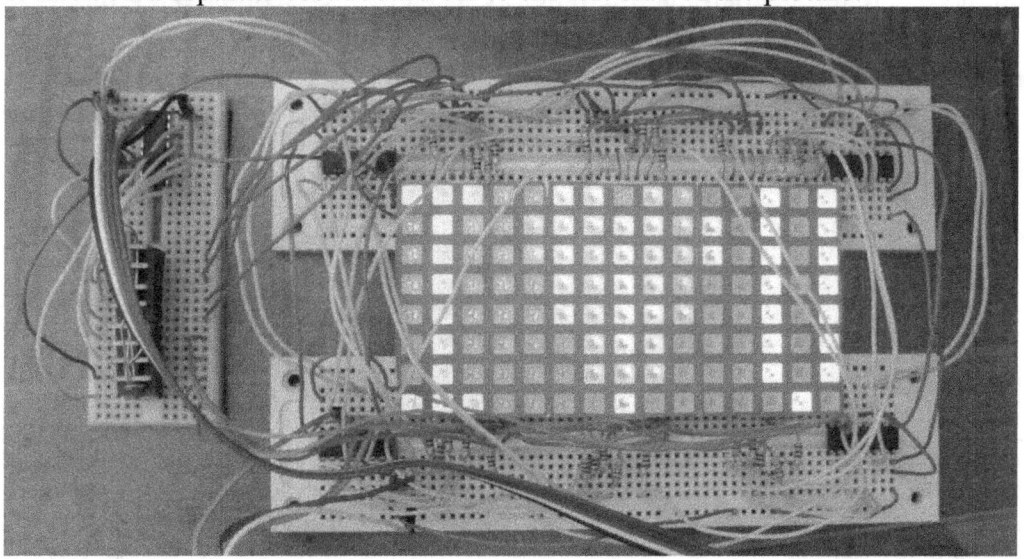

This next picture shows the wiring from another angle. You can clearly see the TIP121 NPN transistor on the left. A screw with a wire wrapped around it connects the tabs together and goes to 5 volts. The base, input is on the left and the emitter, output is on the right.

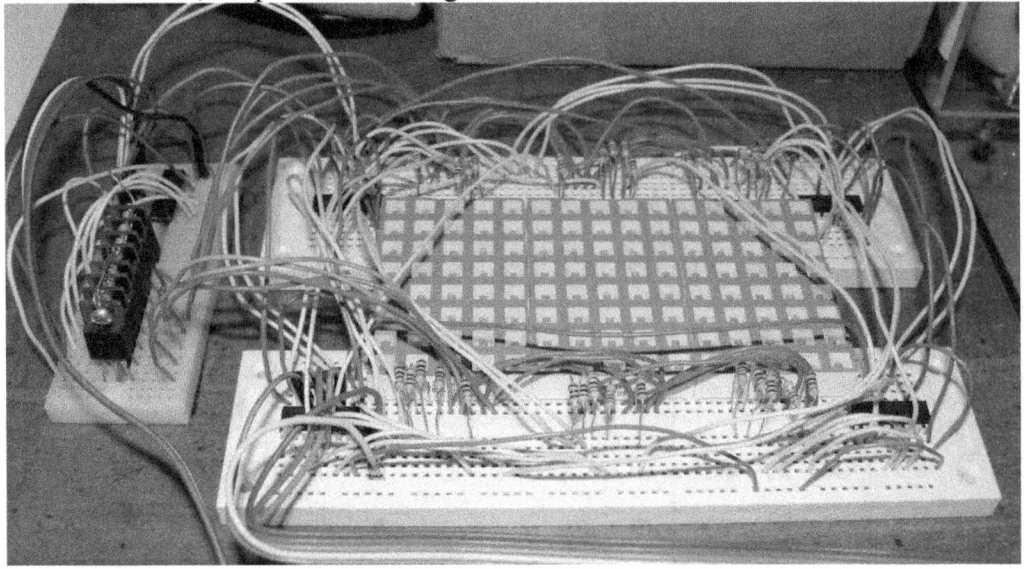

Up next is the schematic diagram of the 8x15 dual color led display. It is not an easy project to build. It can be built one LED array at a time. Later on I added NPN driver transistors between the 74HC595 and the D1-D8 pins.

In the schematic I called the row pins "D" because "R" would be confused with "Red". Note that pins 11 and 12 are common to all shift registers. However for each color pin 9 goes to pin 14 of the next shift register. This wiring daisy chains the 8 bit shift registers together so they behave like 16 bit shift registers.

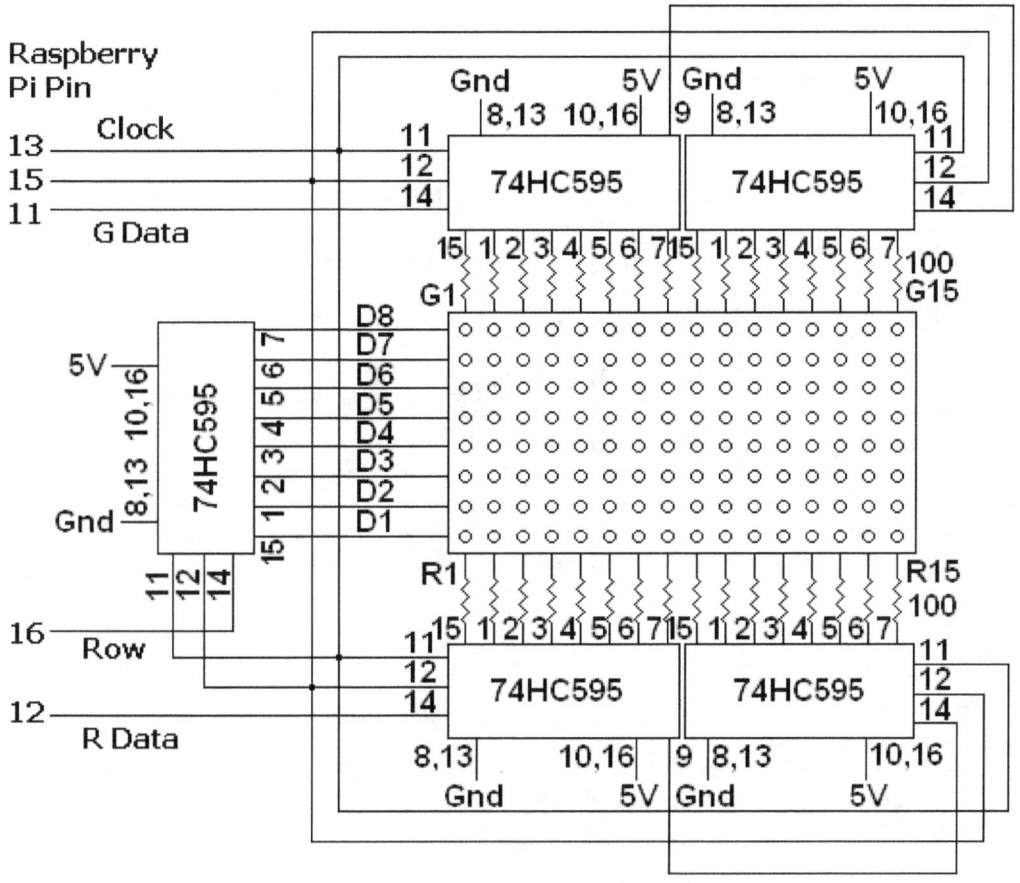

Here is the Python code to make the 8x15 Dual color LED array work. Note the use of a third array to feed the row selection shift register.

```
# LED 8x15 Dual Color array
import RPi.GPIO as GPIO
import time
GPIO.setmode(GPIO.BCM)
GPIO.setup(17, GPIO.OUT) # Red Data
GPIO.setup(18, GPIO.OUT) # Green Data
GPIO.setup(27, GPIO.OUT) # Clock, 21 on older
GPIO.setup(22, GPIO.OUT) # Enable
GPIO.setup(23, GPIO.OUT) # Row Data
GPIO.setup(24, GPIO.OUT)
GPIO.setup(25, GPIO.OUT)

Redbyte1 =[0,0,0,0,0,0,1,1,0,1,1,0,0,1,0,1]
Redbyte2 =[0,0,0,0,0,1,1,1,1,1,1,1,0,1,0,1]
Redbyte3 =[0,0,0,0,0,1,1,1,1,1,1,1,0,1,0,1]
Redbyte4 =[0,0,0,0,0,0,1,1,1,1,1,0,0,1,0,1]
Redbyte5 =[0,0,0,0,0,0,1,1,1,1,1,0,0,1,0,1]
```

```
Redbyte6 =[0,0,0,0,0,0,0,1,1,1,0,0,0,1,0,1]
Redbyte7 =[0,0,0,0,0,0,0,1,1,1,0,0,0,1,0,1]
Redbyte8 =[0,0,0,0,0,0,0,0,1,0,0,0,0,0,1,0]
Greenbyte1 =[0,1,1,1,0,0,0,0,0,0,0,0,0,1,0,1]
Greenbyte2 =[0,0,1,0,0,0,0,0,0,0,0,0,0,1,0,1]
Greenbyte3 =[0,0,1,0,0,0,0,0,0,0,0,0,0,1,0,1]
Greenbyte4 =[0,0,1,0,0,0,0,0,0,0,0,0,0,1,0,1]
Greenbyte5 =[0,0,1,0,0,0,0,0,0,0,0,0,0,1,0,1]
Greenbyte6 =[0,0,1,0,0,0,0,0,0,0,0,0,0,1,0,1]
Greenbyte7 =[0,0,1,0,0,0,0,0,0,0,0,0,0,1,0,1]
Greenbyte8 =[0,1,1,1,0,0,0,0,0,0,0,0,0,0,1,0]
Rowbyte1 =[0,1,0,0,0,0,0,0,0,1,0,0,0,0,0,0]
Rowbyte2 =[0,0,1,0,0,0,0,0,0,0,1,0,0,0,0,0]
Rowbyte3 =[0,0,0,1,0,0,0,0,0,0,0,1,0,0,0,0]
Rowbyte4 =[0,0,0,0,1,0,0,0,0,0,0,0,1,0,0,0]
Rowbyte5 =[0,0,0,0,0,1,0,0,0,0,0,0,0,1,0,0]
Rowbyte6 =[0,0,0,0,0,0,1,0,0,0,0,0,0,0,1,0]
Rowbyte7 =[0,0,0,0,0,0,0,1,0,0,0,0,0,0,0,1,0]
Rowbyte8 =[0,0,0,0,0,0,0,0,1,0,0,0,0,0,0,0,1]

# set up loop
cycle= 0
while cycle < 100000:
  row = 0
  while row < 8:
    row = row+1
    # Send data to the display
    shift = 15
    while shift > 0:
        # determine if bit is set or clear data is inverted
        GPIO.output(17, GPIO.HIGH)
        GPIO.output(18, GPIO.HIGH)
        GPIO.output(23, GPIO.LOW)
        # Send data to shift registers
        if row==1:
            if Redbyte1[shift] == 1:
              GPIO.output(17, GPIO.LOW)
            if Greenbyte1[shift] == 1:
              GPIO.output(18, GPIO.LOW)
            if Rowbyte1[shift] == 1:
              GPIO.output(23, GPIO.HIGH)
        if row==2:
            if Redbyte2[shift] == 1:
              GPIO.output(17, GPIO.LOW)
            if Greenbyte2[shift] == 1:
              GPIO.output(18, GPIO.LOW)
            if Rowbyte2[shift] == 1:
              GPIO.output(23, GPIO.HIGH)
        if row==3:
            if Redbyte3[shift] == 1:
```

```
            GPIO.output(17, GPIO.LOW)
          if Greenbyte3[shift] == 1:
            GPIO.output(18, GPIO.LOW)
          if Rowbyte3[shift] == 1:
            GPIO.output(23, GPIO.HIGH)
        if row==4:
          if Redbyte4[shift] == 1:
            GPIO.output(17, GPIO.LOW)
          if Greenbyte4[shift] == 1:
            GPIO.output(18, GPIO.LOW)
          if Rowbyte4[shift] == 1:
            GPIO.output(23, GPIO.HIGH)
        if row==5:
          if Redbyte5[shift] == 1:
            GPIO.output(17, GPIO.LOW)
          if Greenbyte5[shift] == 1:
            GPIO.output(18, GPIO.LOW)
          if Rowbyte5[shift] == 1:
            GPIO.output(23, GPIO.HIGH)
        if row==6:
          if Redbyte6[shift] == 1:
            GPIO.output(17, GPIO.LOW)
          if Greenbyte6[shift] == 1:
            GPIO.output(18, GPIO.LOW)
          if Rowbyte6[shift] == 1:
            GPIO.output(23, GPIO.HIGH)
        if row==7:
          if Redbyte7[shift] == 1:
            GPIO.output(17, GPIO.LOW)
          if Greenbyte7[shift] == 1:
            GPIO.output(18, GPIO.LOW)
          if Rowbyte7[shift] == 1:
            GPIO.output(23, GPIO.HIGH)
        if row==8:
          if Redbyte8[shift] == 1:
            GPIO.output(17, GPIO.LOW)
          if Greenbyte8[shift] == 1:
            GPIO.output(18, GPIO.LOW)
          if Rowbyte8[shift] == 1:
            GPIO.output(23, GPIO.HIGH)
        # advance the clock
        GPIO.output(27, GPIO.HIGH)
        GPIO.output(27, GPIO.LOW)
        shift=shift-1
    # latch and display the data
    GPIO.output(22, GPIO.LOW)
    GPIO.output(22, GPIO.HIGH)
  cycle=cycle+1
# end
```

Chapter 14
8x8 Three Color LED Array

Recently I was able to get my hands on some three color, as in red, green and blue, eight columns by eight row LED arrays. The model number of this LED array is CRM-2388ARGB-L. It is very popular on eBay. The LED array did not come with any wiring instructions, so I set down with a nine volt battery and a 470 ohm resistor and figured out how it is wired up.

Underneath it there are two rows of 16 pins each. The top row does not have the part number near it. The bottom row, left most pin, is the closest pin to the part number. There is also a number "1" located near this pin. This chart shows what I found the pin out to be.

-----Rows-------				---------------Green-----------------							------Rows-----			
D8	D7	D6	D5	G1	G2	G3	G4	G5	G6	G7	G8	D4	D3	D2	D1
32	31	30	29	28	27	26	25	24	23	22	21	20	19	18	17
	0	0	0	0	0		0		0		0	0			
	0	0	0	0	0		0		0		0	0			
	0	0	0	0	0		0		0		0	0			
	0	0	0	0	0		0		0		0	0			
	0	0	0	0	0		0		0		0	0			
	0	0	0	0	0		0		0		0	0			
	0	0	0	0	0		0		0		0	0			
	0	0	0	0	0		0		0		0	0			
1	2	3	4	5	6	7	8	9	10	11	12	13	14	15	16
B1	B2	B3	B4	B5	B6	B7	B8	R1	R2	R3	R4	R5	R6	R7	R8
----------------Blue-----------------								----------------Red-----------------							

Here is the schematic diagram of a typical 8x8 RGB LED array.

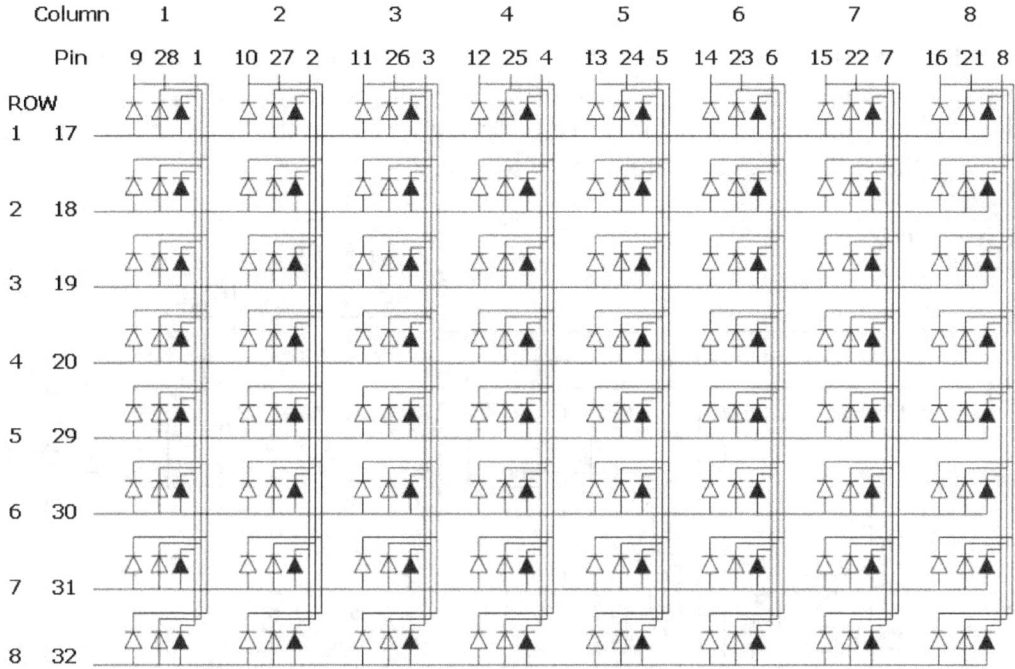

Column   1   2   3   4   5   6   7   8
Pin   9 28 1   10 27 2   11 26 3   12 25 4   13 24 5   14 23 6   15 22 7   16 21 8

ROW
1   17
2   18
3   19
4   20
5   29
6   30
7   31
8   32

Another common three color 8x8 LED display is the NEU-8860-RGB. Here is its pinout as seen from above. The pin connections are somewhat scattered instead of being grouped together.

| R1 | R2 | G2 | R3 | G3 | R4 | D2 | D1 | D3 | G5 | R6 | G6 | D5 | R7 | B8 | R8 |
32	31	30	29	28	27	26	25	24	23	22	21	20	19	18	17
	0		0		0		0		0		0		0		0
	0		0		0		0		0		0		0		0
	0		0		0		0		0		0		0		0
	0		0		0		0		0		0		0		0
	0		0		0		0		0		0		0		0
	0		0		0		0		0		0		0		0
	0		0		0		0		0		0		0		0
	0		0		0		0		0		0		0		0

| 1 | 2 | 3 | 4 | 5 | 6 | 7 | 8 | 9 | 10 | 11 | 12 | 13 | 14 | 15 | 16 |
| G1 | B1 | D7 | B2 | B3 | D8 | B4 | G4 | B5 | R5 | D4 | B6 | D6 | B7 | G7 | G8 |

Next I wired up three 74HC595 shift registers with eight 100 ohm resistors each to the red, green and blue array pins. I was reusing the design that had previously worked for the two color eight by fifteen led array.

For the rows there is another 74HC595 and there could have been some row driver transistors. You might get a little more brightness with the NPN row driver transistors. However it is bright enough without the NPN transistors. Here is the schematic diagram of how to wire it up.

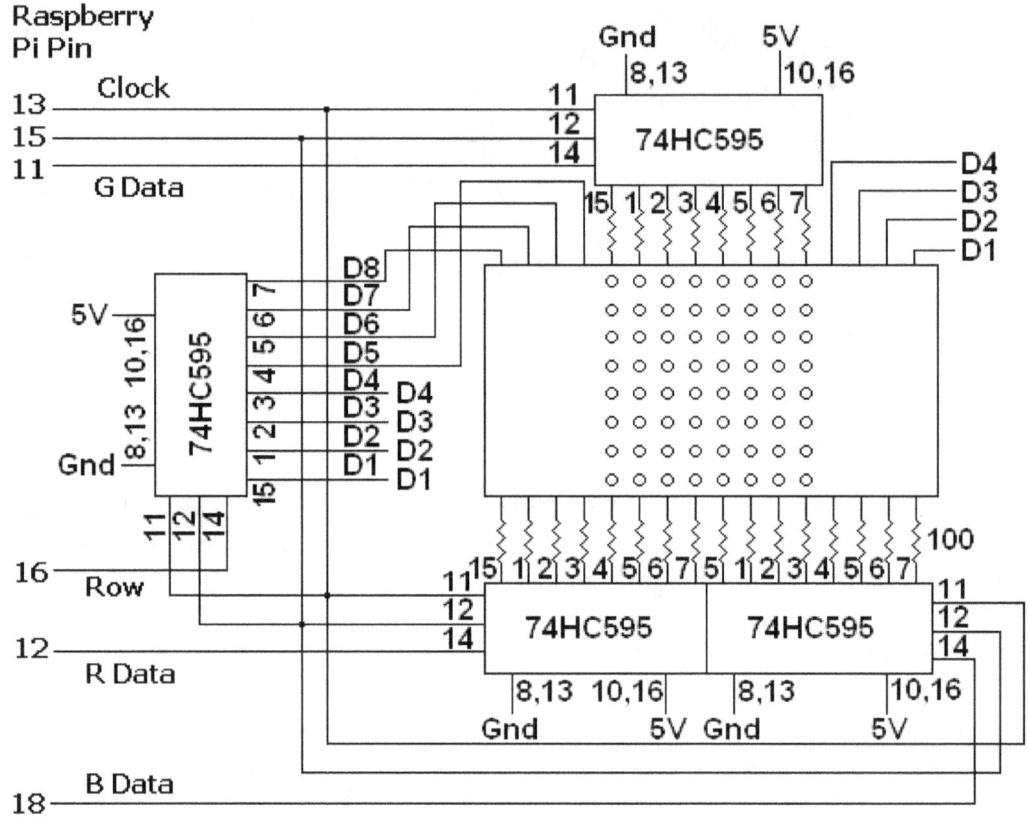

Up next is a picture of the three color 8x8 LED array wired up and working. The top right shift register is the row selecting shift register. Because this book is printed in black and white, here are what the colors are. At the bottom there is green grass, in the middle there is a red car and at the top a blue sky with the sun shining with white and yellow rays.

Here is the Python code to make the car scene pictured above. Note that I got rid of the array for selecting the rows. An additional if statement does the trick of creating the row data.

```
# LED 8x8 RGB Color array
import RPi.GPIO as GPIO
import time
GPIO.setmode(GPIO.BCM)

GPIO.setup(17, GPIO.OUT) # Red Data
GPIO.setup(18, GPIO.OUT) # Green Data
GPIO.setup(27, GPIO.OUT) # Clock, 21 on older
GPIO.setup(22, GPIO.OUT) # Enable
GPIO.setup(23, GPIO.OUT) # Row Data
```

```
GPIO.setup(24, GPIO.OUT) # Blue
GPIO.setup(25, GPIO.OUT) # Not used

Redbyte1 =[0,0,0,0,1,1,1,0]
Redbyte2 =[0,0,0,0,0,1,0,0]
Redbyte3 =[0,0,0,0,0,0,0,0]
Redbyte4 =[0,1,1,1,1,1,0,0]
Redbyte5 =[1,1,1,1,1,1,1,1]
Redbyte6 =[1,0,1,1,1,1,0,1]
Redbyte7 =[0,0,0,0,0,0,0,0]
Redbyte8 =[0,0,0,0,0,0,0,0]

Greenbyte1 =[0,0,0,0,1,1,1,0]
Greenbyte2 =[0,0,0,0,0,1,0,0]
Greenbyte3 =[0,0,0,0,0,0,0,0]
Greenbyte4 =[0,0,0,0,0,0,0,0]
Greenbyte5 =[0,0,0,0,0,0,0,0]
Greenbyte6 =[0,0,0,0,0,0,0,0]
Greenbyte7 =[1,1,1,1,1,1,1,1]
Greenbyte8 =[1,1,1,1,1,1,1,1]

Bluebyte1 =[1,1,1,1,0,1,0,1]
Bluebyte2 =[1,1,1,1,1,0,1,1]
Bluebyte3 =[1,1,1,1,1,1,1,1]
Bluebyte4 =[1,0,0,0,0,0,1,1]
Bluebyte5 =[0,0,0,0,0,0,0,0]
Bluebyte6 =[0,0,0,0,0,0,0,0]
Bluebyte7 =[0,0,0,0,0,0,0,0]
Bluebyte8 =[0,0,0,0,0,0,0,0]

# set up loop
cycle= 0
while cycle < 10000:
  row = 0
  while row < 8:
    row = row+1
    # Send data to the display
    shift = 7
    while shift >= 0:
        # determine if bit is set or clear data is inverted
        GPIO.output(17, GPIO.HIGH)
        GPIO.output(18, GPIO.HIGH)
        GPIO.output(24, GPIO.HIGH)
        GPIO.output(23, GPIO.LOW)
        # Send data to shift registers
        if row==1:
            if Redbyte1[shift] == 1:
              GPIO.output(17, GPIO.LOW)
            if Greenbyte1[shift] == 1:
              GPIO.output(18, GPIO.LOW)
```

```python
        if Bluebyte1[shift] == 1:
          GPIO.output(24, GPIO.LOW)
        if shift == 0:
          GPIO.output(23, GPIO.HIGH)
    if row==2:
        if Redbyte2[shift] == 1:
          GPIO.output(17, GPIO.LOW)
        if Greenbyte2[shift] == 1:
          GPIO.output(18, GPIO.LOW)
        if Bluebyte2[shift] == 1:
          GPIO.output(24, GPIO.LOW)
        if shift == 1:
          GPIO.output(23, GPIO.HIGH)
    if row==3:
        if Redbyte3[shift] == 1:
          GPIO.output(17, GPIO.LOW)
        if Greenbyte3[shift] == 1:
          GPIO.output(18, GPIO.LOW)
        if Bluebyte3[shift] == 1:
          GPIO.output(24, GPIO.LOW)
        if shift == 2:
          GPIO.output(23, GPIO.HIGH)
    if row==4:
        if Redbyte4[shift] == 1:
          GPIO.output(17, GPIO.LOW)
        if Greenbyte4[shift] == 1:
          GPIO.output(18, GPIO.LOW)
        if Bluebyte4[shift] == 1:
          GPIO.output(24, GPIO.LOW)
        if shift == 3:
          GPIO.output(23, GPIO.HIGH)
    if row==5:
        if Redbyte5[shift] == 1:
          GPIO.output(17, GPIO.LOW)
        if Greenbyte5[shift] == 1:
          GPIO.output(18, GPIO.LOW)
        if Bluebyte5[shift] == 1:
          GPIO.output(24, GPIO.LOW)
        if shift == 4:
          GPIO.output(23, GPIO.HIGH)
    if row==6:
        if Redbyte6[shift] == 1:
          GPIO.output(17, GPIO.LOW)
        if Greenbyte6[shift] == 1:
          GPIO.output(18, GPIO.LOW)
        if Bluebyte6[shift] == 1:
          GPIO.output(24, GPIO.LOW)
        if shift == 5:
          GPIO.output(23, GPIO.HIGH)
    if row==7:
```

```python
            if Redbyte7[shift] == 1:
               GPIO.output(17, GPIO.LOW)
            if Greenbyte7[shift] == 1:
               GPIO.output(18, GPIO.LOW)
            if Bluebyte7[shift] == 1:
               GPIO.output(24, GPIO.LOW)
            if shift == 6:
               GPIO.output(23, GPIO.HIGH)
         if row==8:
            if Redbyte8[shift] == 1:
               GPIO.output(17, GPIO.LOW)
            if Greenbyte8[shift] == 1:
               GPIO.output(18, GPIO.LOW)
            if Bluebyte8[shift] == 1:
               GPIO.output(24, GPIO.LOW)
            if shift == 7:
               GPIO.output(23, GPIO.HIGH)
         # Advance the clock
         GPIO.output(27, GPIO.HIGH)
         GPIO.output(27, GPIO.LOW)
         shift=shift-1
      # latch and display the data
      GPIO.output(22, GPIO.LOW)
      GPIO.output(22, GPIO.HIGH)
   cycle=cycle+1
# end
```

Chapter 15
8x40 Dual Color LED Array

A few years back I purchased a lot of "Signature Electronic Signs". They were removed from airports that were upgrading to LCD screens. I interfaced them to an Arduino Uno and sold them with the code to make it work. I still have one laying around and so I interfaced it to the Raspberry Pi. Here is the wiring diagram to make it work with the Raspberry Pi. Pin one is the top left pin.

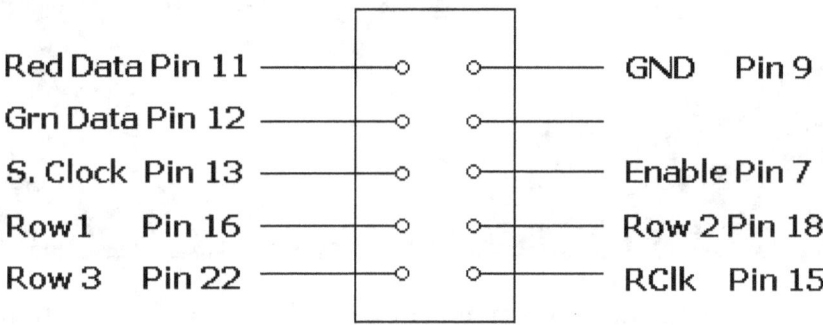

I tried sharing the 5 volt power supply with the Raspberry Pi but that crashed it. It is best to have a separate 5 volt 1 or 2 Amp power supply for running this LED sign.

Here is a picture of the 8x40 LED display in operation. I know Raspberry has two letter R's, but they did not fit on the display.

Coming up is a picture of the actual connector on the sign.

Up next is the schematic diagram, it is turned sideways in order to fit it in this book. The 74138 has some driver transistors with it that I do not have a schematic for. The outputs of the 74138 go through an inverter, and NPN transistor then a PNP power transistor. This setup delivers a full 5 volts to the rows of the LED arrays.

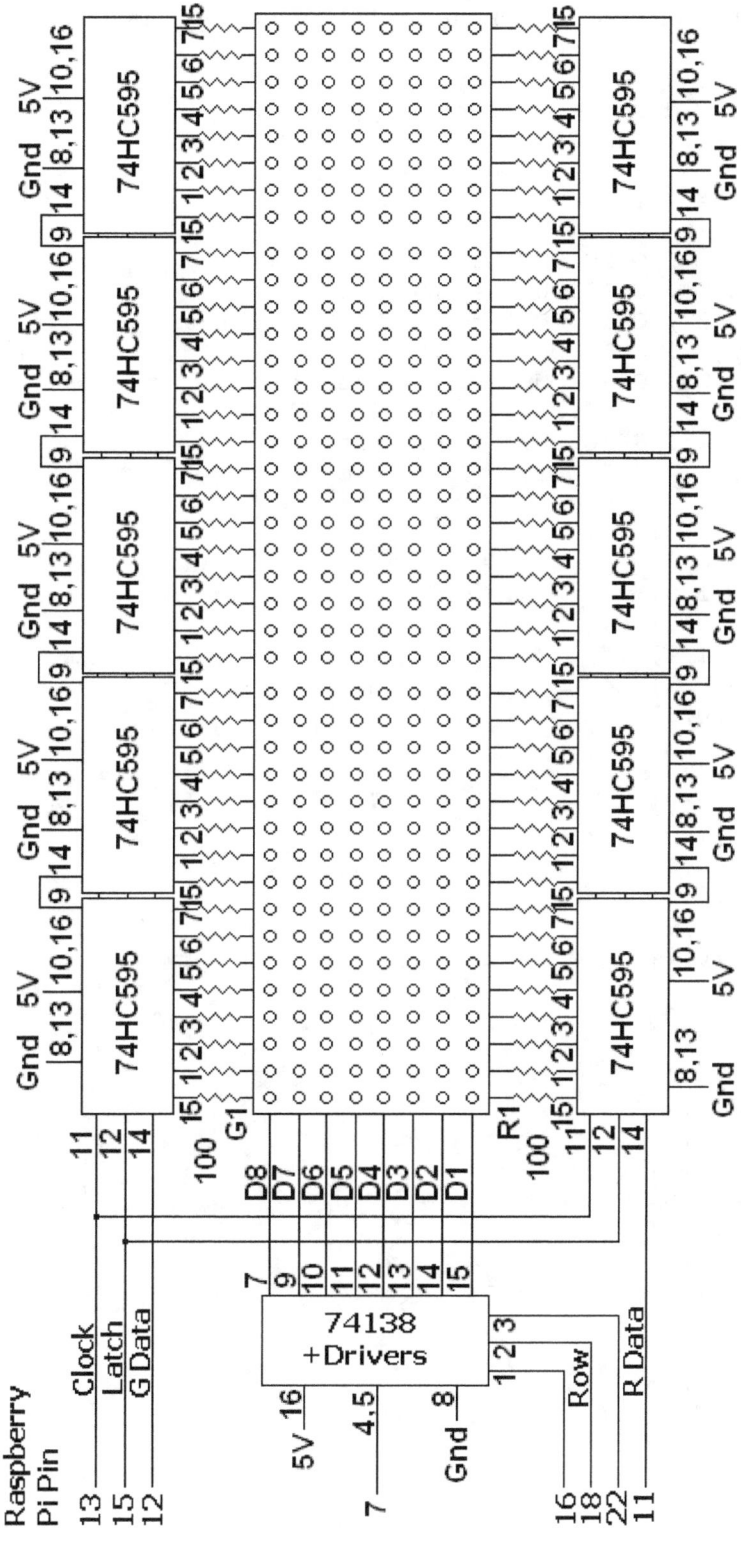

Here is the Python code to make it display something stationary.

```python
# LED 8x40 Dual Color array
# Uses individual letters strung together
import RPi.GPIO as GPIO
import time
GPIO.setmode(GPIO.BCM)

GPIO.setup(17, GPIO.OUT) # Red Data
GPIO.setup(18, GPIO.OUT) # Green Data
GPIO.setup(27, GPIO.OUT) # Clock, 21 on older
GPIO.setup(22, GPIO.OUT) # Latch
GPIO.setup(23, GPIO.OUT) # Row Data 1
GPIO.setup(24, GPIO.OUT) # Row Data 2
GPIO.setup(25, GPIO.OUT) # Row Data 4
GPIO.setup(4, GPIO.OUT) # Row Enable

A1 =[0,0,0,0,0]
A2 =[0,0,1,0,0]
A3 =[0,1,0,1,0]
A4 =[1,0,0,0,1]
A5 =[1,1,1,1,1]
A6 =[1,0,0,0,1]
A7 =[1,0,0,0,1]
A8 =[1,0,0,0,1]

B1 =[0,0,0,0,0]
B2 =[1,1,1,1,0]
B3 =[1,0,0,0,1]
B4 =[1,0,0,0,1]
B5 =[1,1,1,1,0]
B6 =[1,0,0,0,1]
B7 =[1,0,0,0,1]
B8 =[1,1,1,1,0]

E1 =[0,0,0,0,0]
E2 =[1,1,1,1,1]
E3 =[1,0,0,0,0]
E4 =[1,0,0,0,0]
E5 =[1,1,1,1,0]
E6 =[1,0,0,0,0]
E7 =[1,0,0,0,0]
E8 =[1,1,1,1,1]

P1 =[0,0,0,0,0]
P2 =[1,1,1,1,0]
P3 =[1,0,0,0,1]
P4 =[1,0,0,0,1]
P5 =[1,1,1,1,0]
P6 =[1,0,0,0,0]
```

```
P7 =[1,0,0,0,0]
P8 =[1,0,0,0,0]

R1 =[0,0,0,0,0]
R2 =[1,1,1,1,0]
R3 =[1,0,0,0,1]
R4 =[1,0,0,0,1]
R5 =[1,1,1,1,0]
R6 =[1,0,0,0,1]
R7 =[1,0,0,0,1]
R8 =[1,0,0,0,1]

S1 =[0,0,0,0,0]
S2 =[0,1,1,1,1]
S3 =[1,0,0,0,0]
S4 =[1,0,0,0,0]
S5 =[0,1,1,1,0]
S6 =[0,0,0,0,1]
S7 =[0,0,0,0,1]
S8 =[1,1,1,1,0]

Y1 =[0,0,0,0,0]
Y2 =[1,0,0,0,1]
Y3 =[1,0,0,0,1]
Y4 =[0,1,0,1,0]
Y5 =[0,0,1,0,0]
Y6 =[0,0,1,0,0]
Y7 =[0,0,1,0,0]
Y8 =[0,0,1,0,0]

# Z is used as a blank
Z1 =[0,0,0,0,0]
Z2 =[0,0,0,0,0]
Z3 =[0,0,0,0,0]
Z4 =[0,0,0,0,0]
Z5 =[0,0,0,0,0]
Z6 =[0,0,0,0,0]
Z7 =[0,0,0,0,0]
Z8 =[0,0,0,0,0]

Red1=R1+Z1+S1+P1+Z1+E1+R1+Z1
Red2=R2+Z2+S2+P2+Z2+E2+R2+Z2
Red3=R3+Z3+S3+P3+Z3+E3+R3+Z3
Red4=R4+Z4+S4+P4+Z4+E4+R4+Z4
Red5=R5+Z5+S5+P5+Z5+E5+R5+Z5
Red6=R6+Z6+S6+P6+Z6+E6+R6+Z6
Red7=R7+Z7+S7+P7+Z7+E7+R7+Z7
Red8=R8+Z8+S8+P8+Z8+E8+R8+Z8

Green1=R1+A1+Z1+P1+B1+Z1+R1+Y1
```

```
Green2=R2+A2+Z2+P2+B2+Z2+R2+Y2
Green3=R3+A3+Z3+P3+B3+Z3+R3+Y3
Green4=R4+A4+Z4+P4+B4+Z4+R4+Y4
Green5=R5+A5+Z5+P5+B5+Z5+R5+Y5
Green6=R6+A6+Z6+P6+B6+Z6+R6+Y6
Green7=R7+A7+Z7+P7+B7+Z7+R7+Y7
Green8=R8+A8+Z8+P8+B8+Z8+R8+Y8

# set up the loop
cycle= 0
while cycle < 1000:
  row = 0
  while row <9:
    row = row+1
    # Send data to the shift registers
    shift = 39
    while shift >= 0:
        GPIO.output(17, GPIO.LOW)
        GPIO.output(18, GPIO.LOW)
        # determine if bit is set or clear data is NOT
inverted
        if row==1:
          if Red1[shift] == 1: GPIO.output(17, GPIO.HIGH)
          if Green1[shift] == 1: GPIO.output(18, GPIO.HIGH)
        if row==2:
          if Red2[shift] == 1: GPIO.output(17, GPIO.HIGH)
          if Green2[shift] == 1: GPIO.output(18, GPIO.HIGH)
        if row==3:
          if Red3[shift] == 1: GPIO.output(17, GPIO.HIGH)
          if Green3[shift] == 1: GPIO.output(18, GPIO.HIGH)
        if row==4:
          if Red4[shift] == 1: GPIO.output(17, GPIO.HIGH)
          if Green4[shift] == 1: GPIO.output(18, GPIO.HIGH)
        if row==5:
          if Red5[shift] == 1: GPIO.output(17, GPIO.HIGH)
          if Green5[shift] == 1: GPIO.output(18, GPIO.HIGH)
        if row==6:
          if Red6[shift] -- 1: GPIO.output(17, GPIO.HIGH)
          if Green6[shift] == 1: GPIO.output(18, GPIO.HIGH)
        if row==7:
          if Red7[shift] == 1: GPIO.output(17, GPIO.HIGH)
          if Green7[shift] == 1: GPIO.output(18, GPIO.HIGH)
        if row==8:
          if Red8[shift] == 1: GPIO.output(17, GPIO.HIGH)
          if Green8[shift] == 1: GPIO.output(18, GPIO.HIGH)
        # advance the clock
        GPIO.output(27, GPIO.HIGH); GPIO.output(27, GPIO.LOW)
        shift=shift-1
    # select the row data is inverted
    GPIO.output(4, GPIO.HIGH) # Turn off display
```

```
    if row==1: GPIO.output(23, GPIO.HIGH); GPIO.output(24,
GPIO.HIGH); GPIO.output(25, GPIO.HIGH)
    if row==2: GPIO.output(23, GPIO.LOW); GPIO.output(24,
GPIO.HIGH); GPIO.output(25, GPIO.HIGH)
    if row==3: GPIO.output(23, GPIO.HIGH); GPIO.output(24,
GPIO.LOW); GPIO.output(25, GPIO.HIGH)
    if row==4: GPIO.output(23, GPIO.LOW); GPIO.output(24,
GPIO.LOW); GPIO.output(25, GPIO.HIGH)
    if row==5: GPIO.output(23, GPIO.HIGH); GPIO.output(24,
GPIO.HIGH); GPIO.output(25, GPIO.LOW)
    if row==6: GPIO.output(23, GPIO.LOW); GPIO.output(24,
GPIO.HIGH); GPIO.output(25, GPIO.LOW)
    if row==7: GPIO.output(23, GPIO.HIGH); GPIO.output(24,
GPIO.LOW); GPIO.output(25, GPIO.LOW)
    if row==8: GPIO.output(23, GPIO.LOW); GPIO.output(24,
GPIO.LOW); GPIO.output(25, GPIO.LOW)
    # latch and display the data
    GPIO.output(22, GPIO.LOW); GPIO.output(22, GPIO.HIGH)
    GPIO.output(4, GPIO.LOW) # Turn back on
    # time.sleep(.0005)
  cycle=cycle+1
```

Here is another sample of code that makes the sign scroll across the sign.
Note that the characters are now 6 bytes wide, so there is a space to the right
of each character.

```
# LED 8x40 Scroll - Dual Color array
# Uses individual letters strung together
import RPi.GPIO as GPIO
import time
GPIO.setmode(GPIO.BCM)

GPIO.setup(17, GPIO.OUT) # Red Data
GPIO.setup(18, GPIO.OUT) # Green Data
GPIO.setup(27, GPIO.OUT) # Clock, 21 on older
GPIO.setup(22, GPIO.OUT) # Latch
GPIO.setup(23, GPIO.OUT) # Row Data 1
GPIO.setup(24, GPIO.OUT) # Row Data 2
GPIO.setup(25, GPIO.OUT) # Row Data 4
GPIO.setup(4, GPIO.OUT) # Row Enable

A1 =[0,0,0,0,0,0]
A2 =[0,0,1,0,0,0]
A3 =[0,1,0,1,0,0]
A4 =[1,0,0,0,1,0]
A5 =[1,1,1,1,1,0]
A6 =[1,0,0,0,1,0]
A7 =[1,0,0,0,1,0]
A8 =[1,0,0,0,1,0]
```

```
B1 =[0,0,0,0,0,0]
B2 =[1,1,1,1,0,0]
B3 =[1,0,0,0,1,0]
B4 =[1,0,0,0,1,0]
B5 =[1,1,1,1,0,0]
B6 =[1,0,0,0,1,0]
B7 =[1,0,0,0,1,0]
B8 =[1,1,1,1,0,0]

E1 =[0,0,0,0,0,0]
E2 =[1,1,1,1,1,0]
E3 =[1,0,0,0,0,0]
E4 =[1,0,0,0,0,0]
E5 =[1,1,1,1,0,0]
E6 =[1,0,0,0,0,0]
E7 =[1,0,0,0,0,0]
E8 =[1,1,1,1,1,0]

P1 =[0,0,0,0,0,0]
P2 =[1,1,1,1,0,0]
P3 =[1,0,0,0,1,0]
P4 =[1,0,0,0,1,0]
P5 =[1,1,1,1,0,0]
P6 =[1,0,0,0,0,0]
P7 =[1,0,0,0,0,0]
P8 =[1,0,0,0,0,0]

R1 =[0,0,0,0,0,0]
R2 =[1,1,1,1,0,0]
R3 =[1,0,0,0,1,0]
R4 =[1,0,0,0,1,0]
R5 =[1,1,1,1,0,0]
R6 =[1,0,0,0,1,0]
R7 =[1,0,0,0,1,0]
R8 =[1,0,0,0,1,0]

S1 -[0,0,0,0,0,0]
S2 =[0,1,1,1,1,0]
S3 =[1,0,0,0,0,0]
S4 =[1,0,0,0,0,0]
S5 =[0,1,1,1,0,0]
S6 =[0,0,0,0,1,0]
S7 =[0,0,0,0,1,0]
S8 =[1,1,1,1,0,0]

Y1 =[0,0,0,0,0,0]
Y2 =[1,0,0,0,1,0]
Y3 =[1,0,0,0,1,0]
Y4 =[0,1,0,1,0,0]
```

```
Y5 =[0,0,1,0,0,0]
Y6 =[0,0,1,0,0,0]
Y7 =[0,0,1,0,0,0]
Y8 =[0,0,1,0,0,0]

# Z is used as a blank
Z1 = Z2 = Z3 = Z4 = Z5 = Z6 = Z7 = Z8 = [0,0,0,0,0,0]

# pad the arrays with lots of zero's before and after.
Red1=Z1+Z1+Z1+Z1+Z1+Z1+Z1+R1+Z1+S1+P1+Z1+E1+R1+R1+Z1+Z1+Z1+Z1+
Z1+Z1+Z1
Red2=Z1+Z1+Z1+Z1+Z1+Z1+Z1+R2+Z2+S2+P2+Z2+E2+R2+R2+Z2+Z1+Z1+Z1+
Z1+Z1+Z1
Red3=Z1+Z1+Z1+Z1+Z1+Z1+Z1+R3+Z3+S3+P3+Z3+E3+R3+R3+Z3+Z1+Z1+Z1+
Z1+Z1+Z1
Red4=Z1+Z1+Z1+Z1+Z1+Z1+Z1+R4+Z4+S4+P4+Z4+E4+R4+R4+Z4+Z1+Z1+Z1+
Z1+Z1+Z1
Red5=Z1+Z1+Z1+Z1+Z1+Z1+Z1+R5+Z5+S5+P5+Z5+E5+R5+R5+Z5+Z1+Z1+Z1+
Z1+Z1+Z1
Red6=Z1+Z1+Z1+Z1+Z1+Z1+Z1+R6+Z6+S6+P6+Z6+E6+R6+R6+Z6+Z1+Z1+Z1+
Z1+Z1+Z1
Red7=Z1+Z1+Z1+Z1+Z1+Z1+Z1+R7+Z7+S7+P7+Z7+E7+R7+R7+Z7+Z1+Z1+Z1+
Z1+Z1+Z1
Red8=Z1+Z1+Z1+Z1+Z1+Z1+Z1+R8+Z8+S8+P8+Z8+E8+R8+R8+Z8+Z1+Z1+Z1+
Z1+Z1+Z1

Green1=Z1+Z1+Z1+Z1+Z1+Z1+Z1+R1+A1+Z1+P1+B1+Z1+R1+R1+Y1+Z1+Z1+Z
1+Z1+Z1+Z1
Green2=Z1+Z1+Z1+Z1+Z1+Z1+Z1+R2+A2+Z2+P2+B2+Z2+R2+R2+Y2+Z1+Z1+Z
1+Z1+Z1+Z1
Green3=Z1+Z1+Z1+Z1+Z1+Z1+Z1+R3+A3+Z3+P3+B3+Z3+R3+R3+Y3+Z1+Z1+Z
1+Z1+Z1+Z1
Green4=Z1+Z1+Z1+Z1+Z1+Z1+Z1+R4+A4+Z4+P4+B4+Z4+R4+R4+Y4+Z1+Z1+Z
1+Z1+Z1+Z1
Green5=Z1+Z1+Z1+Z1+Z1+Z1+Z1+R5+A5+Z5+P5+B5+Z5+R5+R5+Y5+Z1+Z1+Z
1+Z1+Z1+Z1
Green6=Z1+Z1+Z1+Z1+Z1+Z1+Z1+R6+A6+Z6+P6+B6+Z6+R6+R6+Y6+Z1+Z1+Z
1+Z1+Z1+Z1
Green7=Z1+Z1+Z1+Z1+Z1+Z1+Z1+R7+A7+Z7+P7+B7+Z7+R7+R7+Y7+Z1+Z1+Z
1+Z1+Z1+Z1
Green8=Z1+Z1+Z1+Z1+Z1+Z1+Z1+R8+A8+Z8+P8+B8+Z8+R8+R8+Y8+Z1+Z1+Z
1+Z1+Z1+Z1

# set up the loop
cycle= 0
scroll=0
while cycle < 5000:
  row = 0
  while row <8:
    row = row+1
```

```python
    # Send data to the shift registers
    shift = 39
    while shift >= 0:
        GPIO.output(17, GPIO.LOW)
        GPIO.output(18, GPIO.LOW)
        # determine if bit is set or clear data is NOT
inverted
        if row==0:
          if Red1[shift+scroll] == 1:
            GPIO.output(17, GPIO.HIGH)
          if Green1[shift+scroll] == 1:
            GPIO.output(18, GPIO.HIGH)
        elif row==1:
          if Red2[shift+scroll] == 1:
            GPIO.output(17, GPIO.HIGH)
          if Green2[shift+scroll] == 1:
            GPIO.output(18, GPIO.HIGH)
        elif row==2:
          if Red3[shift+scroll] == 1:
            GPIO.output(17, GPIO.HIGH)
          if Green3[shift+scroll] == 1:
            GPIO.output(18, GPIO.HIGH)
        elif row==3:
          if Red4[shift+scroll] == 1:
            GPIO.output(17, GPIO.HIGH)
          if Green4[shift+scroll] == 1:
            GPIO.output(18, GPIO.HIGH)
        elif row==4:
          if Red5[shift+scroll] == 1:
            GPIO.output(17, GPIO.HIGH)
          if Green5[shift+scroll] == 1:
            GPIO.output(18, GPIO.HIGH)
        elif row==5:
          if Red6[shift+scroll] == 1:
            GPIO.output(17, GPIO.HIGH)
          if Green6[shift+scroll] == 1:
            GPIO.output(18, GPIO.HIGH)
        elif row==6:
          if Red7[shift+scroll] == 1:
            GPIO.output(17, GPIO.HIGH)
          if Green7[shift+scroll] == 1:
            GPIO.output(18, GPIO.HIGH)
        elif row==7:
          if Red8[shift+scroll] == 1:
            GPIO.output(17, GPIO.HIGH)
          if Green8[shift+scroll] == 1:
            GPIO.output(18, GPIO.HIGH)
        # advance the clock
        GPIO.output(27, GPIO.LOW); GPIO.output(27, GPIO.HIGH)
        shift=shift-1
```

```
    # select the row data is inverted
    GPIO.output(4, GPIO.HIGH) # Turn off display
    GPIO.output(23, GPIO.HIGH);
    GPIO.output(24, GPIO.HIGH)
    GPIO.output(25, GPIO.HIGH)
    if row & 0x01: GPIO.output(23, GPIO.LOW)
    if row & 0x02: GPIO.output(24, GPIO.LOW)
    if row & 0x04: GPIO.output(25, GPIO.LOW)
    # latch and display the data
    GPIO.output(22, GPIO.LOW); GPIO.output(22, GPIO.HIGH)
    GPIO.output(4, GPIO.LOW) # Turn display back on
    # time.sleep(.0005)
  cycle=cycle+1
  if (cycle%20)==1:
    scroll=scroll+1
  if scroll > 90: scroll = 0
# end
```

Chapter 16
4x4x5 LED 3D Cube

This is a fairly complex project. To make the cube we must first make four 4x5 LED arrays that are transparent. To do this we will need a wood template with 1/8 inch holes on a one inch by one inch spacing. The 1/8 inch holes are for 3mm LED's. Larger LED's would require larger holes. I used flat top LED's as I had many of them around that were left over from another project. Diffused 3mm LED's should work just as well if not better.

To make the transparent LED arrays first take the longer positive lead of the LED's and bend it into a loop about 3/8 inches from the LED. Cut off the excess lead. Then insert the LED's in the holes that you drilled in the wood. Then take a 5 inch piece of wire and poke it through the loops that you just created. Then solder the wire in place.

The other lead from the LED is bent flat to the wood board. When all five rows of LED's are in place a second wire coming from the other direction is soldered to the second negative lead forming the columns. This is done four times for each of the four columns.

Here is a picture of some LED's with the longer lead bent into a loop.

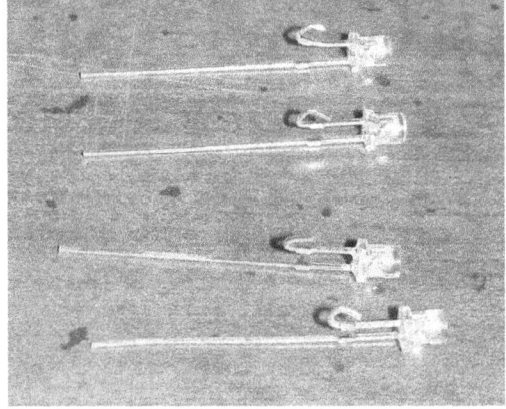

Up next is a picture of the LED's in the wood frame. Note that I marked where to cut the wires so they are all about the same length.

Here is a close up of the LED matrix frame wiring. The left lead is going to
the column wire and the right lead is bent flat and connected to the row wire.

When you have soldered up all 20 LED's you can use a 9 volt battery and a 470 ohm resistor to test out the LED's and then make any repairs that might be necessary. Once verified the LED array can then be removed and then another array can be made, you will need to make four of the arrays.

The arrays should plug into two breadboards on the one inch centers. You might want to install the 100 ohm resistors first. I cut the leads shorter on the resistors too.

Once the arrays are in place you will need to make the cross connectors. I used the same board and inserted round IC socket type connectors in the holes that the LED's were in. Then solder a wire across each 4 of the socket pins to make the cross connectors.

Here is a picture of the cross connectors being made. Also one of the IC sockets is visible that has not been cut into individual pieces yet.

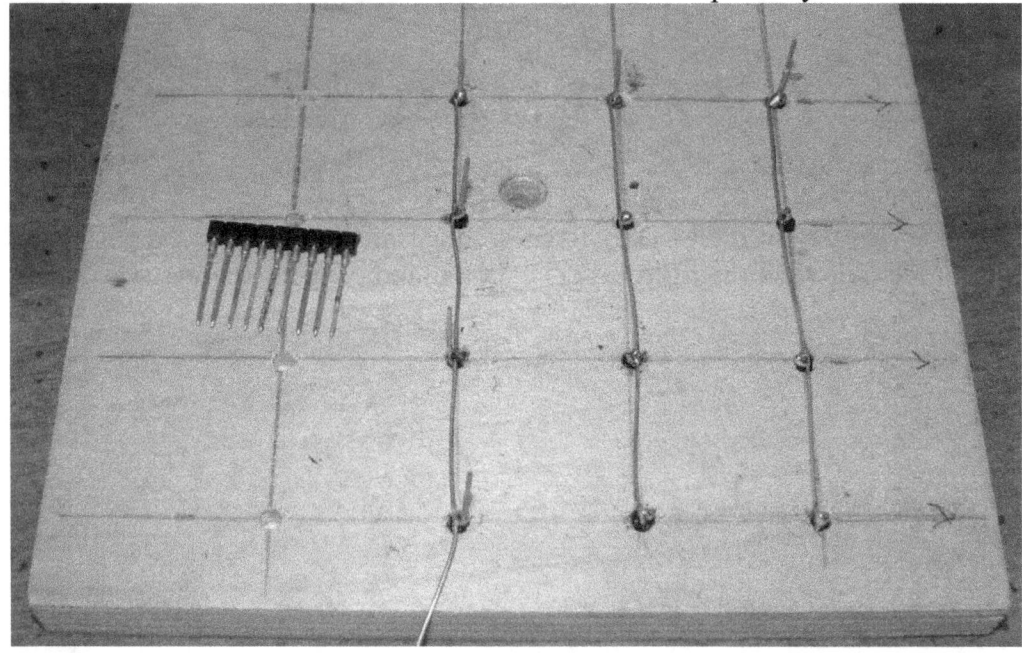

Next add the three IC's and connect them up. I connected pin 7 of the back 74HC595 to the back left corner of the LED arrays. Then connect the other pins following that order. If it is done right the back row of LED's will correspond to the first line of 1's and 0's in the code.

Once it is all wired up run a LED test program. Hopefully you will only damaged one or two LED's and they will need to be replaced if they do not work.

Coming up next is the schematic diagram of the 4x4x5 LED cube. It is once again turned sideways to fit it in this book. The schematic makes it look simpler that it really is to make.

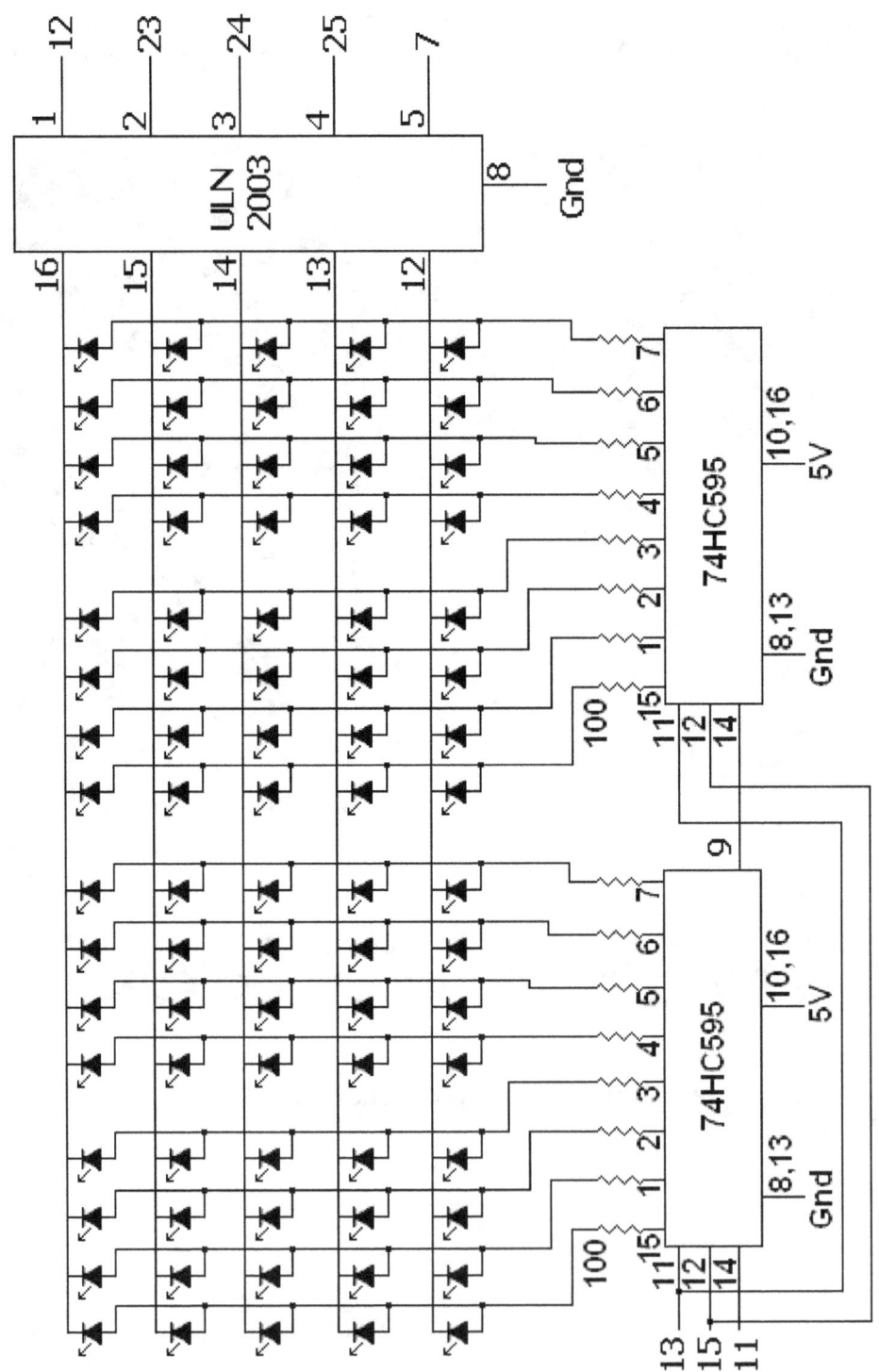

Here is the first example of some code for the 4x4x5 LED cube. This code produces a stationary 3D letter "A".

```
# LED cube A
# Uses 2x74595 and ULN2003
import RPi.GPIO as GPIO
import time
GPIO.setmode(GPIO.BCM)

GPIO.setup(17, GPIO.OUT)  # Serial Data
GPIO.setup(18, GPIO.OUT)  # L1
GPIO.setup(27, GPIO.OUT)  # Clock, 21 on older
GPIO.setup(22, GPIO.OUT)  # Latch
GPIO.setup(23, GPIO.OUT)  # L2
GPIO.setup(24, GPIO.OUT)  # L3
GPIO.setup(25, GPIO.OUT)  # L4
GPIO.setup(4,  GPIO.OUT)  # L5

L1 =[0,1,1,0,0,1,1,0,0,1,1,0,0,1,1,0]
L2 =[1,0,0,1,1,0,0,1,1,0,0,1,1,0,0,1]
L3 =[1,1,1,1,1,1,1,1,1,1,1,1,1,1,1,1]
L4 =[1,0,0,1,1,0,0,1,1,0,0,1,1,0,0,1]
L5 =[1,0,0,1,1,0,0,1,1,0,0,1,1,0,0,1]
R1=L1
R2=L2
R3=L3
R4=L4
R5=L5

# set up the loop
cycle= 0
while cycle < 10000:
  level = 1
  while level < 6:
    shift=0
    while shift < 16:
      # Send data to the shift register
      GPIO.output(17, GPIO.LOW)
      if level ==1:
        if R1[shift] == 1: GPIO.output(17, GPIO.HIGH)
      if level ==2:
        if R2[shift] == 1: GPIO.output(17, GPIO.HIGH)
      if level ==3:
        if R3[shift] == 1: GPIO.output(17, GPIO.HIGH)
      if level ==4:
        if R4[shift] == 1: GPIO.output(17, GPIO.HIGH)
      if level ==5:
        if R5[shift] == 1: GPIO.output(17, GPIO.HIGH)
      # advance the clock
      GPIO.output(27, GPIO.HIGH)
```

```
        GPIO.output(27, GPIO.LOW)
        shift=shift+1
    # turn off the display
    GPIO.output(18, GPIO.LOW)
    GPIO.output(23, GPIO.LOW)
    GPIO.output(24, GPIO.LOW)
    GPIO.output(25, GPIO.LOW)
    GPIO.output(4, GPIO.LOW)
    # latch and display the data
    GPIO.output(22, GPIO.HIGH)
    GPIO.output(22, GPIO.LOW)
    if level == 1: GPIO.output(18, GPIO.HIGH)
    if level == 2: GPIO.output(23, GPIO.HIGH)
    if level == 3: GPIO.output(24, GPIO.HIGH)
    if level == 4: GPIO.output(25, GPIO.HIGH)
    if level == 5: GPIO.output(4, GPIO.HIGH)
    time.sleep(.002)
    level = level+1
  cycle=cycle+1
# end
```

This is the second example of some code for the 4x4x5 LED cube. It produces an animated display with up to 10 changing steps. This example uses only 5 of those steps.

```
# LED cube - Animated display
# Uses 2x74595 and ULN2003
import RPi.GPIO as GPIO
import time
GPIO.setmode(GPIO.BCM)

GPIO.setup(17, GPIO.OUT) # Serial Data
GPIO.setup(18, GPIO.OUT) # L1
GPIO.setup(27, GPIO.OUT) # Clock, 21 on older
GPIO.setup(22, GPIO.OUT) # Latch
GPIO.setup(23, GPIO.OUT) # L2
GPIO.sctup(24, GPIO.OUT) # L3
GPIO.setup(25, GPIO.OUT) # L4
GPIO.setup(4,  GPIO.OUT) # L5

L1 =[1,0,0,0,0,1,0,0,0,0,1,0,0,0,0,1]
L2 =[0,1,0,0,0,0,1,0,0,0,0,1,0,0,0,0]
L3 =[0,0,1,0,0,0,0,1,0,0,0,0,1,0,0,0]
L4 =[0,0,0,1,0,0,0,0,1,0,0,0,0,1,0,0]
L5 =[0,0,0,0,1,0,0,0,0,1,0,0,0,0,1,0]
R1=L1
R2=L2
R3=L3
R4=L4
```

```
R5=L5

# set up the loop
cycle= 0
while cycle < 10000:
  level = 1
  while level < 6:
    shift=0
    while shift < 16:
      # Send data to the shift register
      GPIO.output(17, GPIO.LOW)
      if level ==1:
        if R1[shift] == 1: GPIO.output(17, GPIO.HIGH)
      if level ==2:
        if R2[shift] == 1: GPIO.output(17, GPIO.HIGH)
      if level ==3:
        if R3[shift] == 1: GPIO.output(17, GPIO.HIGH)
      if level ==4:
        if R4[shift] == 1: GPIO.output(17, GPIO.HIGH)
      if level ==5:
        if R5[shift] == 1: GPIO.output(17, GPIO.HIGH)
      # advance the clock
      GPIO.output(27, GPIO.HIGH)
      GPIO.output(27, GPIO.LOW)
      shift=shift+1
    # turn off display
    GPIO.output(18, GPIO.LOW)
    GPIO.output(23, GPIO.LOW)
    GPIO.output(24, GPIO.LOW)
    GPIO.output(25, GPIO.LOW)
    GPIO.output(4, GPIO.LOW)
    # latch and display the data
    GPIO.output(22, GPIO.HIGH)
    GPIO.output(22, GPIO.LOW)
    if level == 1: GPIO.output(18, GPIO.HIGH)
    if level == 2: GPIO.output(23, GPIO.HIGH)
    if level == 3: GPIO.output(24, GPIO.HIGH)
    if level == 4: GPIO.output(25, GPIO.HIGH)
    if level == 5: GPIO.output(4, GPIO.HIGH)
    time.sleep(.002)
    level = level+1
  # change up the display
  if (cycle/20%10 == 1):
    R1=L1;    R2=L2;    R3=L3;    R4=L4;    R5=L5
  if (cycle/20%10 == 2):
    R1=L2;    R2=L3;    R3=L4;    R4=L5;    R5=L1
  if (cycle/20%10 == 3):
    R1=L3;    R2=L4;    R3=L5;    R4=L1;    R5=L2
  if (cycle/20%10 == 4):
    R1=L4;    R2=L5;    R3=L1;    R4=L2;    R5=L3
```

```
    if (cycle/20%10 == 5):
      R1=L5;      R2=L1;      R3=L2;      R4=L3;      R5=L4
    if (cycle/20%10 == 6):
      R1=L1;      R2=L2;      R3=L3;      R4=L4;      R5=L5
    if (cycle/20%10 == 7):
      R1=L2;      R2=L3;      R3=L4;      R4=L5;      R5=L1
    if (cycle/20%10 == 8):
      R1=L3;      R2=L4;      R3=L5;      R4=L1;      R5=L2
    if (cycle/20%10 == 9):
      R1=L4;      R2=L5;      R3=L1;      R4=L2;      R5=L3
    if (cycle/20%10 == 0):
      R1=L5;      R2=L1;      R3=L2;      R4=L3;      R5=L4
    cycle=cycle+1
# end
```

Chapter 17
8x8x8 LED 3D Cube

For this next project we are going to make a cube that is four times as complex. First the spacing between the LED's was reduced from one inch to .6 inches. That way the LED leads can be soldered between the LED's without adding any extra wires. In fact I cut about ¼ inch off the longer lead.

First you will need a piece of wood at least five inches by five inches in size. Lay out an 8x8 cross grid that is at 6/10 of an inch intervals. If you go any bigger then you will need to mount the shift registers on a separate breadboard. This spacing is also compatible with a common 8x8x8 LED kit as seen on Instructables. The kit says that the LED spacing is 1.5 cm. Drill 1/8 inch diameter holes for 3mm LED's.

On each LED bend the shorter or negative lead at about 1/4 inch or just above the funny spot in the lead. Put 8 of them into the template. Bend the long leads down in the other direction, and solder them together. Then do the same for the next 8 LED's but this time also solder the shorter leads together.

On the next page there is a picture of the board with the first row of LED's in place.

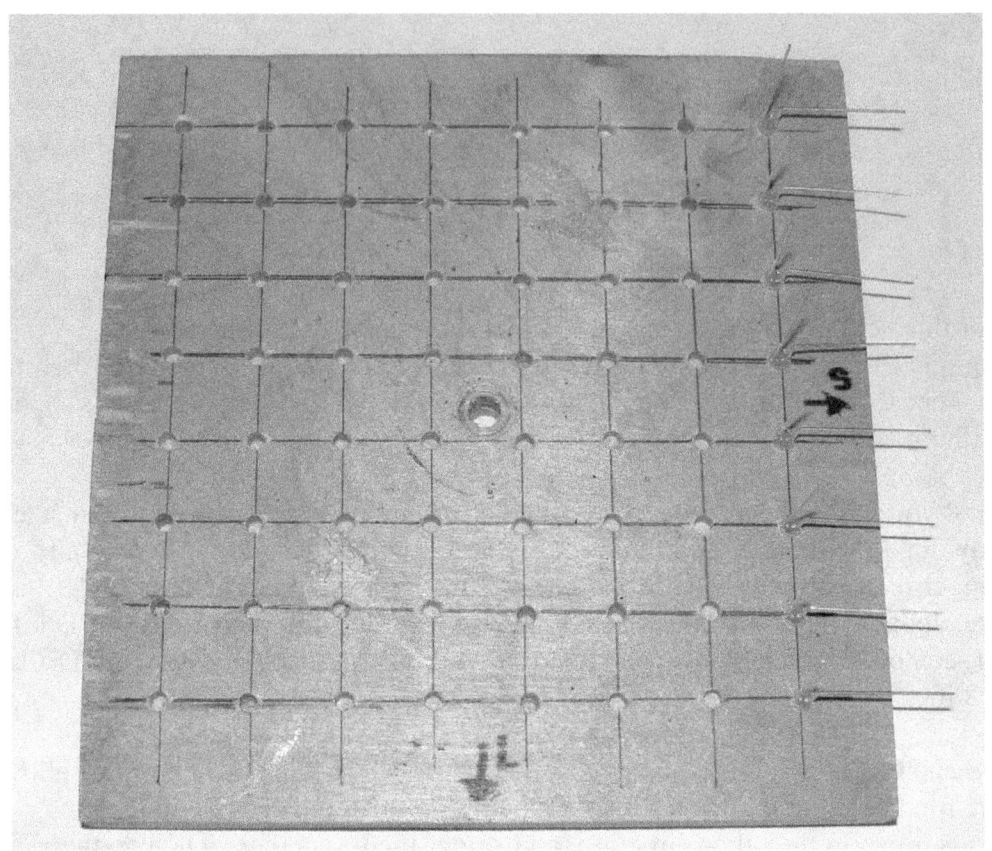

The next picture shows four rows that have been completed. If you install the LED's early you will have to remove them in order to bend their leads.

Here is another close up view from the other direction so you can see how the LED's are connected. The negative, shorter leads are about to be soldered.

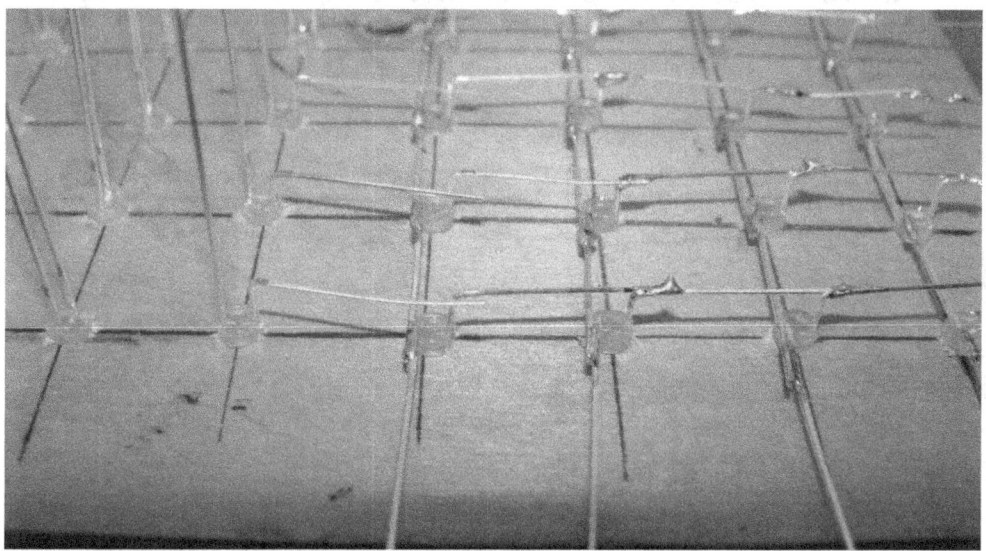

The next picture is of a finished 8x8 LED array. Once all 64 LED's are soldered test them out with a 9 volt battery and a 470 ohm resistor. Also it is a bit tricky to remove the array from the template. You have to gently pull up on each section until it finally lifts out of the wood template.

You will also need to modify the breadboards for this project. The tighter spacing requires that the breadboards be closer together than one inch. To accomplish that we will need to remove the power strips from both sides of the breadboard. These instructions apply to the MB-102 as well as many other breadboards. You will need four of these breadboards to fit the eight 8x8 LED arrays at two per breadboard.

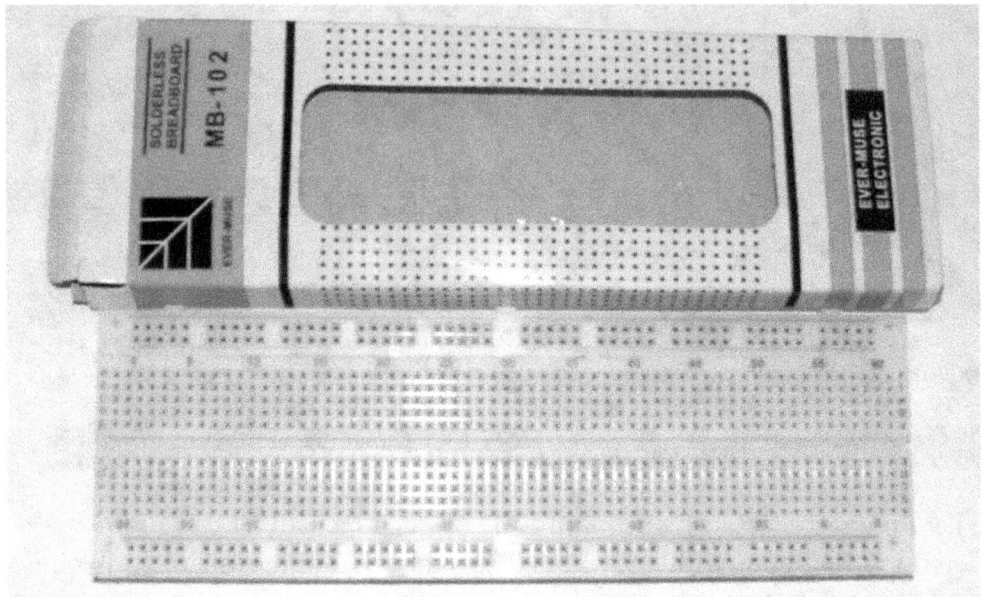

Flip the breadboard over and cut the backing 3/8 of an inch in from each side. The power strips then should either slide up or down to remove them.

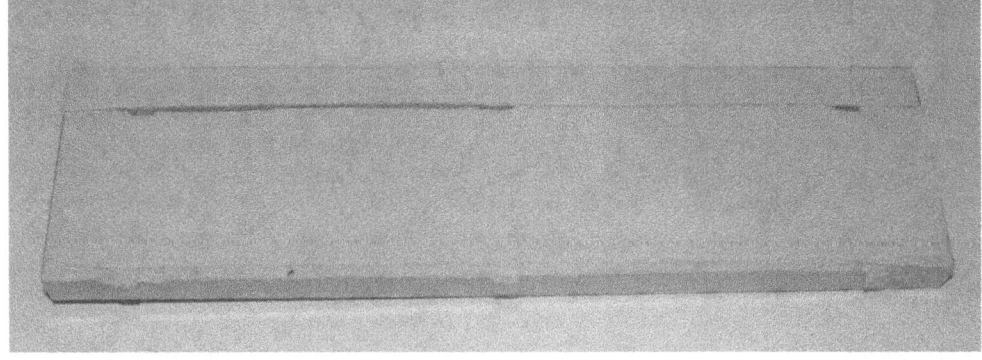

Once the power strips are removed you can usually connect them together. The spacing should now be about .7 inches. That is a little more than the .6 inch spacing between the LED's but it will work. You can now start adding the LED arrays, adding the two 74HC595's, and then the 100 ohm resistors. Up next there is a picture that shows what you should have assembled so far.

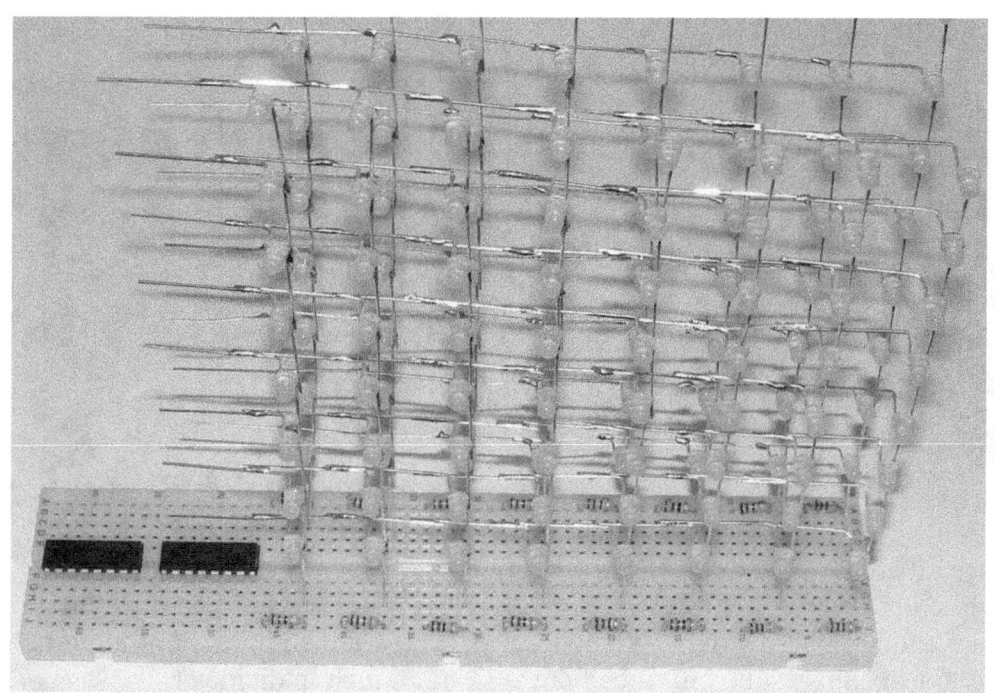

Next you add the 16 jumper wires to the resistors. You will need some two inch, four inch and six inch jumpers. The furthest resistor goes to pin 15, then the next goes to pin one, etc. The right IC goes to the back array and the left IC goes to the front array. Up next is a picture with the 16 jumpers installed.

Note that the centers of the arrays need to be slightly offset on the outside. The breadboards are at .7 inch centers and the arrays are only at .6 inches.

You will need to use row E, I, C, H, C, H, B and F as seen in the next picture. One of my breadboards did not match the other three.

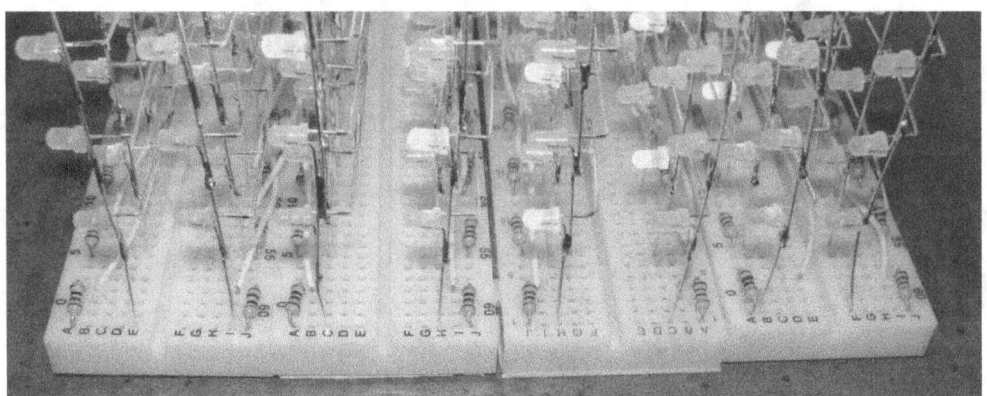

Once all eight of the 8x8 LED arrays are made then you will need to make eight cross connects. I used the same board as was used for the arrays but using another board would be a good idea. The 1/8 inch holes have too much play for the pins as the pins are about 1/16 of an inch in diameter. It was really hard to get the pins straight in the larger holes. I included an IC socket and some salvaged pins on the right side of this picture. To get the pins out of the socket cut the socket close to the pins and then pry the pin out of the socket.

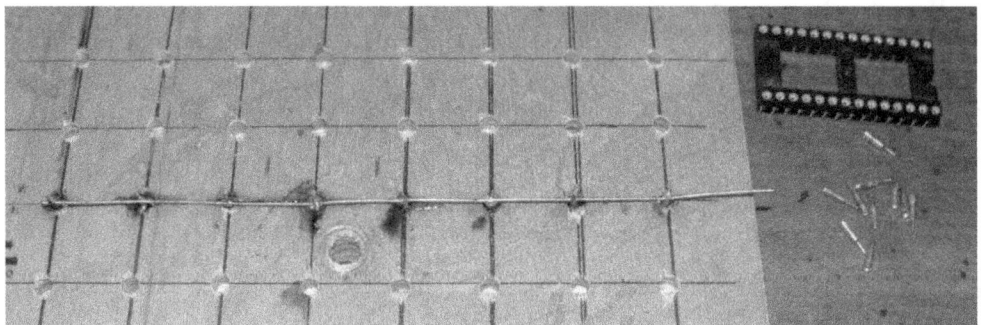

Coming up next is a picture of the completed 8x8x8 LED cube running a LED test. The wiring is fairly difficult for the inner IC's as they are located underneath of the array interconnects. If you connect the wire from pin 14 of the first 74HCT595 to 5 volts then all of the LED's should come on to test them to see if they all work.

On the next page there is the schematic diagram for two of the eight cube layers. The ULN2803 IC is shared among all eight of the cube's layers. Each 8x8 LED array has its own 74595 shift register. So on each breadboard there should be two arrays on the right and two shift registers on the left. Four modules like what is shown in the schematic diagram are made, then the four modules are connected together to make the 8x8x8 LED cube.

Note that if all 64 of the LED's are lit at the same time at 10 ma each you would have 640 ma and thus exceed the ULN2803 power rating. This did not cause any problems in any of my tests, but it is very rare to have all of the LED's on at the same time.

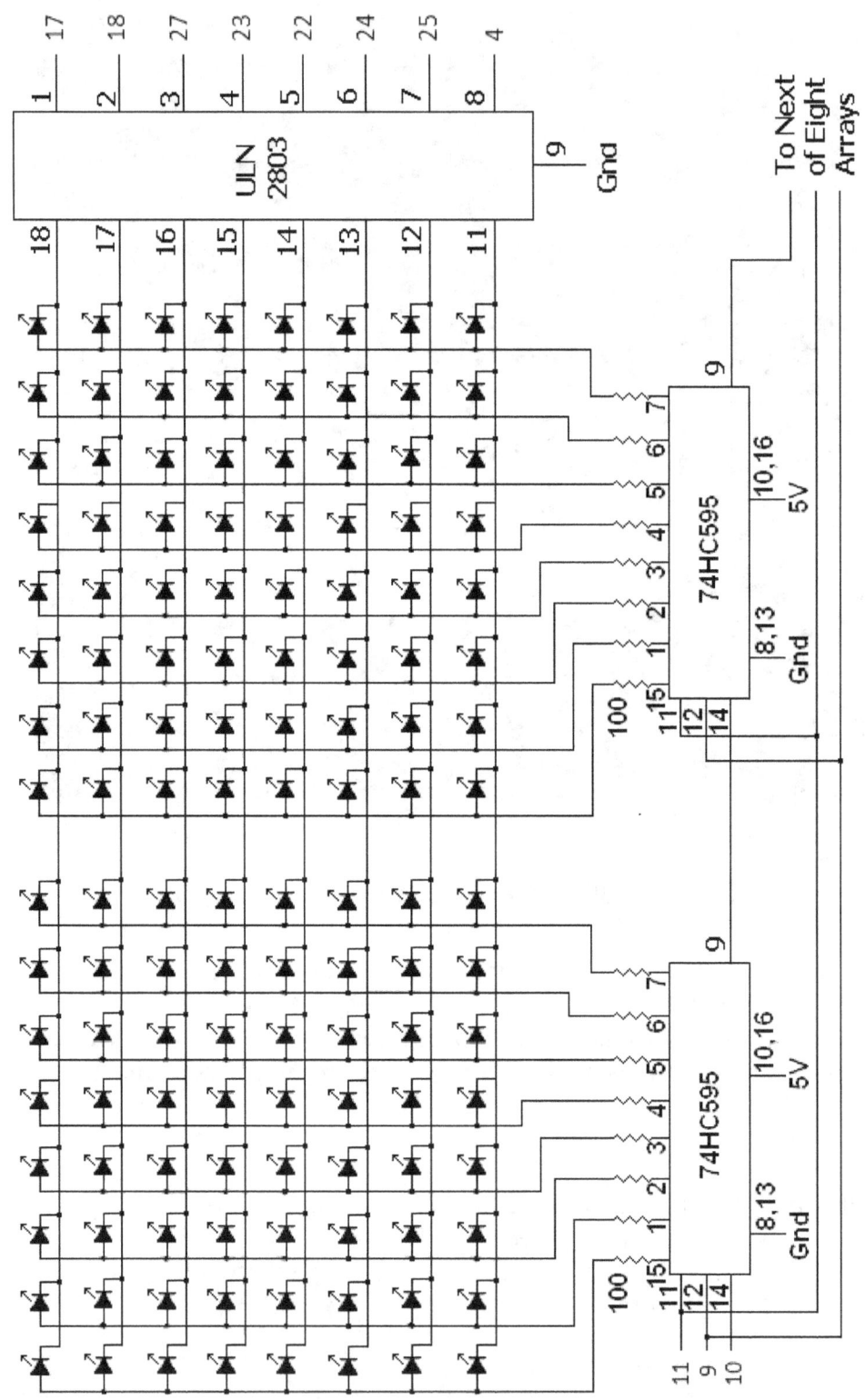

Here is the Python code to make it work.

```python
# 8x8x8 cube random demo

import RPi.GPIO as GPIO
import time
import random
GPIO.setmode(GPIO.BCM)

GPIO.setup(17, GPIO.OUT) # layer 1
GPIO.setup(18, GPIO.OUT) # layer 2
GPIO.setup(27, GPIO.OUT) # layer 3
GPIO.setup(22, GPIO.OUT) # layer 4
GPIO.setup(23, GPIO.OUT) # layer 5
GPIO.setup(24, GPIO.OUT) # layer 6
GPIO.setup(25, GPIO.OUT) # layer 7
GPIO.setup(4, GPIO.OUT) # layer 8
GPIO.setup(9, GPIO.OUT) # latch
GPIO.setup(10, GPIO.OUT) # data
GPIO.setup(11, GPIO.OUT) # clock

# set up loop
cycle= 0
while cycle < 100000:
  level = 0
  while level < 8:
    # Send data to the display
    shift = 0
    while shift < 64:
        # determine if bit is set or clear data is
inverted
        GPIO.output(10, GPIO.LOW)
        rnum = random.randrange(0, 64, 1)
      # Send data to shift registers
      if rnum == shift: GPIO.output(10, GPIO.HIGH)
      # advance the clock
      GPIO.output(11, GPIO.HIGH)
      GPIO.output(11, GPIO.LOW)
      shift=shift+1
    # turn off display
    GPIO.output(17, GPIO.LOW)
    GPIO.output(18, GPIO.LOW)
    GPIO.output(27, GPIO.LOW)
    GPIO.output(22, GPIO.LOW)
    GPIO.output(23, GPIO.LOW)
    GPIO.output(24, GPIO.LOW)
```

```
      GPIO.output(25, GPIO.LOW)
      GPIO.output(4, GPIO.LOW)
      # latch and display the data
      GPIO.output(9, GPIO.HIGH)
      GPIO.output(9, GPIO.LOW)
      if level == 0:  GPIO.output(17, GPIO.HIGH)
      if level == 1:  GPIO.output(18, GPIO.HIGH)
      if level == 2:  GPIO.output(27, GPIO.HIGH)
      if level == 3:  GPIO.output(22, GPIO.HIGH)
      if level == 4:  GPIO.output(23, GPIO.HIGH)
      if level == 5:  GPIO.output(24, GPIO.HIGH)
      if level == 6:  GPIO.output(25, GPIO.HIGH)
      if level == 7:  GPIO.output(4, GPIO.HIGH)
      level = level+1
      time.sleep(.01)
   cycle=cycle+1
```

This next program produces a falling rain effect.

```
# 8x8x8 cube random rain demo

import RPi.GPIO as GPIO
GPIO.setmode(GPIO.BCM)
import time
import random

GPIO.setup(17, GPIO.OUT) # layer 1
GPIO.setup(18, GPIO.OUT) # layer 2
GPIO.setup(27, GPIO.OUT) # layer 3
GPIO.setup(22, GPIO.OUT) # layer 4
GPIO.setup(23, GPIO.OUT) # layer 5
GPIO.setup(24, GPIO.OUT) # layer 6
GPIO.setup(25, GPIO.OUT) # layer 7
GPIO.setup(4, GPIO.OUT) # layer 8
GPIO.setup(9, GPIO.OUT) # latch
GPIO.setup(10, GPIO.OUT) # data
GPIO.setup(11, GPIO.OUT) # clock

# create random numbers
rnum1 = random.randrange(0, 64, 1)
rnum2 = random.randrange(0, 64, 1)
rnum3 = random.randrange(0, 64, 1)
rnum4 = random.randrange(0, 64, 1)
rnum5 = random.randrange(0, 64, 1)
rnum6 = random.randrange(0, 64, 1)
rnum7 = random.randrange(0, 64, 1)
```

```
rnum8 = random.randrange(0, 64, 1)

# set up loop
cycle= 0
while cycle < 100000:
  level = 0
  while level < 8:
    # Send data to the arrays
    shift = 0
    while shift < 64:
      # determine if bit is set or clear
      GPIO.output(10, GPIO.LOW)
      # Send data to the shift registers
      if level == 0:
        if rnum1 == shift: GPIO.output(10, GPIO.HIGH)
      if level == 1:
        if rnum2 == shift: GPIO.output(10, GPIO.HIGH)
      if level == 2:
        if rnum3 == shift: GPIO.output(10, GPIO.HIGH)
      if level == 3:
        if rnum4 == shift: GPIO.output(10, GPIO.HIGH)
      if level == 4:
        if rnum5 == shift: GPIO.output(10, GPIO.HIGH)
      if level == 5:
        if rnum6 == shift: GPIO.output(10, GPIO.HIGH)
      if level == 6:
        if rnum7 == shift: GPIO.output(10, GPIO.HIGH)
      if level == 7:
        if rnum8 == shift: GPIO.output(10, GPIO.HIGH)
      # advance the clock
      GPIO.output(11, GPIO.HIGH)
      GPIO.output(11, GPIO.LOW)
      shift=shift+1
    # turn off display
    GPIO.output(17, GPIO.LOW)
    GPIO.output(18, GPIO.LOW)
    GPIO.output(27, GPIO.LOW)
    GPIO.output(22, GPIO.LOW)
    GPIO.output(23, GPIO.LOW)
    GPIO.output(24, GPIO.LOW)
    GPIO.output(25, GPIO.LOW)
    GPIO.output(4, GPIO.LOW)
    # latch and display the data
    GPIO.output(9, GPIO.HIGH)
    GPIO.output(9, GPIO.LOW)
    if level == 0:  GPIO.output(17, GPIO.HIGH)
```

```
    if level == 1:  GPIO.output(18, GPIO.HIGH)
    if level == 2:  GPIO.output(27, GPIO.HIGH)
    if level == 3:  GPIO.output(22, GPIO.HIGH)
    if level == 4:  GPIO.output(23, GPIO.HIGH)
    if level == 5:  GPIO.output(24, GPIO.HIGH)
    if level == 6:  GPIO.output(25, GPIO.HIGH)
    if level == 7:  GPIO.output(4, GPIO.HIGH)
    level = level+1
# shift random numbers
if cycle%10 == 1 or cycle%10 == 5:
    rnum8 = rnum7
    rnum7 = rnum6
    rnum6 = rnum5
    rnum5 = rnum4
    rnum4 = rnum3
    rnum3 = rnum2
    rnum2 = rnum1
    rnum1 = random.randrange(0, 64, 1)
cycle=cycle+1
```

Bibliography

These are some books that I used the most in writing this book.

Raspberry Pi
A Quick-Start Guide
Maik Schmidt
The Pragmatic Bookshelf
Dallas, Texas • Raleigh, North Carolina
Copyright © 2012 The Pragmatic Programmers, LLC.

Python Tutorial
Simply Easy Learning by tutorialspoint.com
Copyright © tutorialspoint.com

Raspberry Pi Cookbook
by Simon Monk
Copyright © 2014 Simon Monk. All rights reserved.

Here are some really great web sites:

This web site is a tremendous resource for the Raspberry Pi.
http://www.raspberrypi.org/

Adafruit has a special section for the Raspberry Pi.
https://learn.adafruit.com/category/raspberry-pi

There are many free magazines that you can download.
http://www.themagpi.com/issues/